THE PHYSICS OF FLOW
THROUGH POROUS MEDIA

THE PHYSICS OF FLOW THROUGH POROUS MEDIA

By

ADRIAN E. SCHEIDEGGER, Ph.D.

Geophysicist,
Dominion Observatory, Ottawa

THE MACMILLAN COMPANY
NEW YORK
1957

COPYRIGHT 1957
BY UNIVERSITY OF TORONTO PRESS
ALL RIGHTS RESERVED

REPRINTED 1958

CHEMISTRY

PRINTED IN GREAT BRITAIN

To
JAMES W. YOUNG

PREFACE

THE present monograph is not the outcome of a project undertaken in order to produce a book. Rather, the book grew on its own account out of the need felt by research workers to obtain an idea of the present state of knowledge about the physical principles of hydrodynamics in porous media.

If one scans through the literature on porous media, it soon becomes evident that a vast amount of information has been scattered over a multitude of journals. No attempts have ever been made to collect or to co-ordinate this information. The only existing text-books on flow through porous media are either rather out of date or restricted to special aspects of the subject. They are thus unsuitable for the research worker.

The writer, while employed by Imperial Oil Limited in Calgary, was asked to collect as complete a bibliography on the subject as was possible. It was first thought that this would be a relatively easy task, but the list of references grew larger and larger until about two thousand papers had been collected. It became evident that a simple bibliography would be utterly inadequate as a guide to such an enormous amount of literature. The papers were therefore sifted and an attempt was made to relate in a coherent manner the essence of the information contained in them.

It was obviously impossible to present *all* the pertinent information on the hydrodynamics in porous media in detail except by writing a ten-volume encyclopaedia on the subject. Thus a selection had to be made. The guiding principles were as follows: (1) Emphasis was laid on the general physical aspects of the phenomena rather than on particular cases applicable, say, to special engineering problems. (2) Of the many solutions available for some of the basic differential equations, only *one* was chosen for presentation in each case. The theory of differential equations is a well-established discipline of mathematics and has been considered of interest in the present context only if pertinent physical concepts were revealed. (3) The theoretical aspects have perhaps been stressed somewhat more than the experimental ones. However, descriptions of such procedures which enable one to determine theoretical "constants" have always been supplied in order to establish the proper logical sequence.

The above-mentioned principles for selecting material for presentation in the present monograph are admittedly somewhat arbitrary and naturally reflect the writer's own preference. Nevertheless, it is thought that sufficient care has been taken to supply enough literature references from the two thousand that had been originally collected to enable anyone interested in any particular phase of the subject to obtain the required information from the original sources.

The presentation of material contained in this book often follows closely the presentation employed in the original sources. Wherever applicable, this has been indicated in the text by stating that the exposition in a certain section is "following a certain author." It is hoped that no oversights in such acknowledgments have occurred although this is quite possible in view of the fact that the annotated bibliography, upon which this monograph is based, was originally not intended for publication. Backtracing of the provenance of information after the years during which this book had been a confidential report, was sometimes difficult to achieve and the writer wishes to offer his apologies if any credits which are due, should have been omitted. The co-operation of the American Institute of Mining and Metallurgical Engineers, the American Chemical Society, the American Society of Mechanical Engineers, the Soil Science Society of America, the United States Bureau of Mines, the United States National Bureau of Standards, the Danish Academy of Technical Sciences, the Royal Society of London, and the Editor of the journal *Research* (London) in permitting the extensive use of information contained in their respective journals, often more or less verbatim, is hereby equally gratefully acknowledged. The writer is also indebted to the *Petroleum Engineer*, to the *Producers Monthly*, to the *Journal of Applied Physics*, to *Geofisica Pura e Applicata*, and to the *Handbuch der Physik* published by Springer in Berlin, for permission to use material from his own articles which had been published previously. Finally, Mr. R. D. Dymond of the Engineering Laboratory of Imperial Oil Limited in Calgary, Dr. H. L. McDermot of the National Defence Chemical Laboratory in Ottawa, and Dr. L. B. Lerner of the Gulf Research and Development Company in Pittsburgh have read and criticized parts of the manuscript and the writer wishes to express his appreciation for their help.

Publication has been made possible by the release of the original bibliography which was compiled by the writer for Imperial Oil Limited, and by the award of a generous grant-in-aid by the same firm towards the printing costs of the present book. The Canadian Government has contributed time of the writer's to the completion of the study, and the University of Toronto Press has been most helpful in effecting the publication. The writer wishes to thank all these institutions for their splendid co-operation.

The book is dedicated to Mr. James W. Young who, as manager of the Technical Service Department of Imperial Oil Limited, has done so much to encourage basic research in the oil industry in Canada. Without his continued support and understanding, the writer would have had no chance even to attempt the present study. Its appearance now in print is therefore largely due to his efforts. It is the intention that the present dedication should give expression to the writer's deep indebtedness and sincere gratitude.

A. E. S.

Ottawa, Ontario
July 28, 1956

CONTENTS

		PAGE
	Preface	vii
	Introduction	3

PART I. POROUS MEDIA

CHAP.
1.1	Description and Geometrical Properties of Porous Media	5
1.2	The Measurement of Porosity	8
1.2.1	Methods	8
1.2.2	Specific Cases	11
1.3	The Measurement of Specific Surface	11
1.3.1	Methods	11
1.3.2	Specific Cases	14
1.4	The Determination of Pore Size Distribution	14
1.5	The Correlation between Grain Size and Pore Structure	15
1.5.1	The Measurement of Grain Size	15
1.5.2	Theory of Packing of Spheres	16
1.5.3	Packing of Natural Materials	19
1.6	Rheological Properties of Porous Media	20

PART II: FLUIDS

2.1	General Remarks	23
2.2	Continuous Matter Theory	23
2.2.1	The Fundamental Equations	23
2.2.2	Special Cases of Viscous Fluid Flow	24
2.3	Molecular Flow	26
2.4	The Interaction of Fluids with Surfaces	28
2.4.1	Adsorption	28
2.4.2	Interfacial Tension	28
2.5	Miscible Fluids	31

PART III: HYDROSTATICS IN POROUS MEDIA

3.1	Principles of Hydrostatics	36
3.2	Adsorption of Fluids by Porous Media	36

CHAP.		PAGE
3.2.1	Theory of Isotherms	36
3.2.2	Applications of Isotherms to Area Measurements	39
3.3	Capillary Condensation of Fluids in Porous Media	41
3.3.1	Theory of Sorption Hysteresis	41
3.3.2	Determinations of Pore Size Distribution	42
3.4	The Quasistatic Displacement of One Liquid by Another in a Porous Medium	43
3.4.1	Theory of Capillary Pressure	43
3.4.2	Experiments on Capillary Pressure	48
3.4.3	Capillary Pressure and Pore Size Distribution	49
3.5	Relative Wettability	52

PART IV: DARCY'S LAW

4.1	The Flow of Homogeneous Fluids in Porous Media	54
4.2	Experimental Investigations	55
4.2.1	The Experiment of Darcy	55
4.2.2	The Permeability Concept	57
4.3	Differential Forms of Darcy's Law	60
4.3.1	Isotropic Porous Media	60
4.3.2	Compressible Porous Media	62
4.3.3	Anisotropic Porous Media	63
4.4	The Measurement of Permeability	66
4.5	Filtration Theory	68

PART V: SOLUTIONS OF DARCY'S LAW

5.1	General Remarks	70
5.2	Steady State Flow	71
5.2.1	Analytical Solutions	71
5.2.2	Solutions by Analogies	75
5.3	Gravity Flow with a Free Surface	77
5.3.1	Physical Aspects of the Phenomenon	77
5.3.2	Specific Solutions	80
5.4	Unsteady State Flow	82
5.4.1	Unsteady State Flow in Linear Approximation	82

		PAGE
5.4.2	General Unsteady State Flow	85
5.4.3	Experimental Solutions	88

PART VI: PHYSICAL ASPECTS OF PERMEABILITY

6.1	Empirical Correlations	89
6.2	Capillaric Models	91
6.2.1	The Concept of Models	91
6.2.2	Straight Capillaric Model	92
6.2.3	Parallel Type Models	94
6.2.4	Serial Type Models	95
6.2.5	Comparison with Tests	98
6.3	Hydraulic Radius Theories	99
6.3.1	Principles of Theory	99
6.3.2	The Kozeny Theory	100
6.3.3	Modifications of the Kozeny Theory	104
6.3.4	Experiments in Connection with the Hydraulic Radius Theory	105
6.4	Structure Determinations of Porous Media	106
6.4.1	Determination of Surface Area	106
6.4.2	Determination of Other Geometrical Quantities	106
6.5	Criticism of Hydraulic Radius Theory; Other Theories	107
6.5.1	Criticism of the Kozeny Theory	107
6.5.2	Drag Theories of Permeability	108
6.5.3	Statistical Theories	113
6.5.4	Analogy of Laminar Flow in Porous Media with Turbulent Flow in Bulk Fluids	120

PART VII: GENERAL FLOW EQUATIONS

7.1	Limitations of Darcy's Law	123
7.1.1	High Flow Rates	123
7.1.2	Molecular Effects	125
7.1.3	Other Effects	126
7.2	Equations for High Flow Velocities	127
7.2.1	Heuristic Correlations	127
7.2.2	Theoretical Equations	133
7.2.3	Solutions of Turbulent Flow Equations	140
7.3	Equations Accounting for Molecular Effects	144

CHAP.		PAGE
7.3.1	Gas Slippage and Molecular Streaming	144
7.3.2	Adsorption and Diffusion	148
7.4	Applications of General Flow Equations	152
7.4.1	Permeability Corrections	152
7.4.2	Structure Determinations of Porous Media	152

PART VIII: MULTIPLE PHASE FLOW

8.1	General Remarks	153
8.2	Laminar Flow of Immiscible Fluids	154
8.2.1	Qualitative Investigations	154
8.2.2	Darcy's Law	154
8.2.3	Measurement of Relative Permeability	157
8.3	Solutions of Darcy's Law (Immiscible Fluids)	163
8.3.1	The Buckley-Leverett Case	163
8.3.2	Other Analytical Solutions; Elementary Treatment	167
8.3.3	Imbibition	171
8.3.4	Solutions Subject to Constraints	173
8.4	Scaling and Experimental Investigations into Flow Patterns of Immiscible Fluids in Porous Media	176
8.4.1	Dynamical Similarity in Porous Media	176
8.4.2	Experimental Displacement Studies	180
8.5	Physical Aspects of Relative Permeability	181
8.5.1	Visual Studies	181
8.5.2	Theoretical Studies	182
8.6	General Flow Equations for Immiscible Fluids	190
8.6.1	Limitations of Darcy's Law	190
8.6.2	Turbulence	191
8.6.3	Molecular Effects	197
8.7	Miscible Displacement	197
8.7.1	General Principles	197
8.7.2	Specific Calculations	199
8.7.3	Microscopic Theories of Miscible Displacement	200
	List of Symbols	203
	Bibliography	206
	Index	233

THE PHYSICS OF FLOW
THROUGH POROUS MEDIA

INTRODUCTION

THE study of the physics of flow through porous media has become basic for many scientific and engineering applications, quite apart from the interest it holds for its purely scientific aspects. Such diversified fields as soil mechanics, ground water hydrology, petroleum engineering, water purification, industrial filtration, ceramic engineering, powder metallurgy, and the study of gas masks all rely heavily upon it as fundamental to their individual problems. All these branches of science and engineering have contributed a vast amount of literature on the subject.

The scope of hydromechanics in porous media, quite naturally, allows for division into several parts. First, there are the properties of *porous media* and of *fluids*, by themselves, which have to be discussed. The next step is to proceed to the interactions between the two, first to the *statics*, then to the *dynamics*.

The hydro*dynamics* in porous media, in turn, can be classified into hydrodynamics of *single* phase fluids, and hydrodynamics of *several* fluids. Most applications of the flow through porous media are concerned with one single fluid only and thus there is much more information on single phase flow available than on multiple phase flow. Therefore a large part of the book will be devoted to the flow of homogeneous, single phase fluids through porous media and only a smaller part to multiple phase flow.

Most of the information presented in this monograph is collected from original articles in journals. The writer is aware of only three text-books on the subject and these are either out of date or else very restricted in their scope. The first is a book by Muskat (1937) first published in 1937 and reprinted without any changes, it appears, in 1946. It therefore does not account for the developments in the more recent years. The second is a very concise book by Leïbenzon (1947) which gives an excellent exposition of the physical principles of flow through porous media up to 1947. Unfortunately, it does not give many references to the original articles in which the various equations were developed although it mentions the names of their inventors. As a guide to literature and a lead to further scientific study it is therefore rather restricted. Finally, the third text-book is a monograph by Polubarinova-Kochina (1952) which gives a multitude of finite (integral) solutions of the differential equations of flow through porous media for various external conditions. However, it deals almost exclusively with solutions of the *laminar* differential flow equations, and thus does not discuss phenomena deviating from the simplest physical case. It is generally concerned with the mathematics of solving the Laplace equation rather than with the actual physics of the flow phenomena, and this makes it of somewhat limited interest.

Apart from the text-books mentioned above, there are some general reviews that appeared in journals and "Handbooks" which cover part of the subject of flow through porous media. Thus, a lengthy article by Zimens (1944) discusses the characteristics of porous media; a similar work by Engelund (1953) reviews certain aspects of flow through porous media, and, finally, an article of my own (Scheidegger, 1956) gives a brief review of the subject.

Pertinent information has also been collected from text-books that have been written in connection with various engineering applications already mentioned, such as petroleum production, soil mechanics, and so on. None of these text-books, however, even attempt to present a coherent exposition of the physical principles of flow through porous media.

It can be seen that there is really no compilation of the subject-matter of this monograph in existence which would be suitable for the research worker, and it will be found that the present compilation is an almost entirely new presentation of this material.

PART I

POROUS MEDIA

1.1. Description and Geometrical Properties of Porous Media

In order to study the flow of fluids through porous media it is first of all necessary to clarify what is understood by the terms that denote the two materials involved: "fluids" and "porous media."

Starting with the latter, one may be tempted to define "porous media" as solid bodies that contain "pores," it being assumed as intuitively quite clear what is meant by a "pore." However, it is unfortunately much more difficult to give an exact geometrical definition of what is meant by the notion of a "pore" than may appear at first glance. A special effort must therefore be made to obtain a proper description.

Intuitively, "pores" are void spaces which must be distributed more or less frequently through the material if the latter is to be called "porous." Extremely small voids in a solid are called "molecular interstices," very large ones are called "caverns." "Pores" are void spaces intermediate between caverns and molecular interstices; the limitation of their size is therefore intuitive and rather indefinite.

The pores in a porous system may be *interconnected* or *non-interconnected*. Flow of intersticial fluid is possible only if at least part of the pore space is interconnected. The interconnected part of the pore system is called the *effective* pore space of the porous medium.

According to the above description (the term "definition" would hardly be applicable), the following things are examples of porous media: towers packed with pebbles, Berl saddles, Raschig rings, etc.; beds formed of sand, granules, lead shot, etc.; porous rocks such as limestone, pumice, dolomite, etc.; fibrous aggregates such as cloth, felt, filter paper, etc.; and finally catalytic particles containing extremely fine "micro"-pores.

It is thus seen that the term "porous media" encompasses a very wide variety of substances. Because of this, it is desirable to arrange porous media into several classes according to the types of pore spaces which they contain. Any one porous medium, of course, is not restricted to only one class of pore spaces and thus may have pore spaces belonging to several classes. A suitable series of classes of pore spaces was devised by Manegold (1937, 1941) by dividing pore spaces into voids, capillaries, and force spaces: *voids* are characterized by the fact that their walls have only an insignificant effect upon hydrodynamical phenomena in their interior, in *capillaries*, the walls do have a significant effect upon hydrodynamical phenomena in their interior, but do not, however, bring the molecular structure of the fluid into evidence, in *force spaces* the molecular structure of the fluid is brought into evidence.

In addition, pore spaces have also been classified according to whether they are *ordered*, or *disordered*; and also according to whether they are *dispersed* (as in beds of particles) or *connected*. The meaning of these terms is self-evident.

Thus, a porous medium is characterized by a variety of geometrical properties. First of all, the fraction of void to total volume is important. This fraction is called *porosity* (denoted in this monograph by P) and is expressed either as a fraction of 1 or in per cent. If the calculation of porosity is based upon the interconnected pore space instead of on the total pore space, the resulting value is termed *effective* porosity.

Second, another well-defined geometrical quantity of a porous medium is its *specific internal area*. This is the ratio of internal area to bulk volume and it is therefore expressed as a reciprocal length. In this monograph, it will always be denoted by S.

It would be most desirable to be able to define a geometrical quantity that would characterize the *size of the pores* in any one porous medium. Unfortunately, the pore system of a porous body forms a very complicated surface, geometrically difficult to describe. It does not seem to be possible to find any simple parameter or any simple function that would describe this surface. It is, of course, theoretically possible to give the analytical equation of it, but the practical difficulties are quite unsurmountable.

The determination of the pore structure, therefore, is necessarily a somewhat arbitrary undertaking, depending on what property of that surface one wants to describe; it will never be possible to encompass *all* its properties.

Intuitively, one would like to talk about the "size" of pores, a convenient measure of the "size" being the "diameter." However, the term "diameter" makes sense geometrically only if all the pores are of spherical shape—unless some further specifications are attached. If one has the flow of fluids through those pores in mind, however, it will not do to restrict oneself to spherical pores only (the latter would have no effective porosity at all); one will visualize the pores as rather tubeshaped things. One would then call the diameter of such a tube "pore diameter." Unfortunately, even this visualization of the term "diameter" is geometrically quite meaningless unless the tubes are circular. In general they will not be circular, and (which makes matters worse) they will not even possess a normal cross-section since the walls will be irregularly diverging and converging. Thus one cannot even speak about a "biggest" or a "smallest" diameter of the tube at any one point.

The writer is not aware of any citation in the literature where this geometrical dilemma would have been satisfactorily solved. Nevertheless, people talk about the "size" of pores, "pore size distribution," etc., without defining accurately what this is supposed to mean. A possible way out of the dilemma might be by defining the pore diameter (denoted in this monograph by δ) at any one point within the pore space as the diameter of the largest sphere which contains this point and remains wholly within the pore space. Thus, to each point of the pore space a "diameter" could be attached rigorously, and if desired, a "pore size distribution" could be defined—simply by determining what fraction α of the total pore space has a "pore diameter" between δ and $\delta + d\delta$. It is easily seen that the pore size distribution thus defined is normalized in the following way:

$$\int_{0}^{\infty} \alpha(\delta) d\delta = 1 \qquad (1.1.1)$$

Instead of dealing with α, it will often be more convenient to deal with the cumulative pore size distribution, which will be denoted here by f; the latter is defined as that fraction of the pore space which has a pore diameter larger than δ:

$$f(\delta) = \int_{\delta}^{\infty} \alpha(\delta) d\delta; \quad f(0) = 1 \qquad (1.1.2)$$

Once "pore size distribution" is defined, one can try to characterize it by certain parameters. This is best accomplished by fitting various standard distribution curves known from mathematical statistics to actual pore size distribution curves; the parameters of the standard distribution curves will then serve as parameters for the pore size distribution curves. The simplest standard distribution curves suitable for this purpose are the various types of Gauss curves and they have been thus employed by Tollenaar and Blockhuis (1950).

In contrast to pore size distribution, it is easier to determine *grain size* or grain size distribution of a porous medium, at least if the latter is unconsolidated (dispersed). Unfortunately, the grain size distribution as such does not mean too much as far as the properties of the corresponding pore space are concerned. In order to obtain correlations between grain size and pore size, one has first to investigate the theory of packing of grains (cf. sec. 1.5.2).

Finally, another geometrical property has been proposed which has been called *tortuosity*. In this monograph, it will be denoted by T. Originally it was introduced as a kinematical property equal to the relative average length of the flow path of a fluid particle from one side of the porous medium to the other. It is thus a dimensionless quantity. If a suitable model of the porous medium is chosen, viz., one consisting of capillaries, then the "tortuosity" also becomes a purely geometrical property. It could be measured, for example, by an electrical resistivity measurement, as the current would have only one set of paths to flow through which would be identical to the flow paths of fluid particles, namely, the paths prescribed by the capillaries making up the porous body. This idea has been followed up by Fricke (1931), Stamm (1931), Archie (1942), and by Wyllie and Spangler (1952). However, Owen (1952) seems to have shown that there is no direct correlation between electrical and geometrical properties of a porous medium. The notion of tortuosity is, therefore, somewhat doubtful.

1.2. The Measurement of Porosity

1.2.1. Methods.
The measurement of the geometrical quantity defined as "porosity" in section 1.1 can be achieved by a great variety of methods. All these methods aim to measure somehow the void volume and the bulk volume of the material; the ratio of the two, then, is the porosity. The following is a review of the more common of the possible methods.

A. Direct method. The most direct method of determining the porosity is to measure the bulk volume of a piece of porous material and then to compact the body so as to destroy all its voids, and to measure the difference in volumes. Unfortunately, this can be done only if the body is very soft. This method, therefore, has not been applied very extensively, but it is very suitable for the analysis of bread (Lȳnovskiĭ and Postnikova, 1940).

B. Optical methods. Another direct way to determine the porosity is simply by looking at a section of the porous medium with a microscope. The reason why a value of porosity can be obtained in this manner is that the plane porosity of a random section must be the same as that of the porous medium. It might frequently be advantageous first to photograph the section through the microscope and then to planimeter it. Such visual methods of porosity determination have been applied by Zavodovskaya (1937) to porcelain, by Dallmann (1941) to bread, and by Verbeck (1947) to concrete. A stochastic method for the evaluation of a photomicrograph as to porosity has been described by Chalkley, Cornfield and Park (1949): a pin is thrown into the picture, and if its point is in a void area a "hit" is scored. If the experiment is repeated often enough, the ratio of "hits" to "throws" is equal to the porosity of the specimen.

It is not always possible to make sections of a porous medium conveniently. Difficulties will be encountered especially if the porous medium is dispersed. Techniques have therefore been developed whereby the medium is first impregnated with wax or plastics. As a matter of convenience, a dye can be added to the impregnating material so as to make the voids more visible. Such methods have been described by Waldo and Yuster (1937), Imbt and Ellison (1946), Nuss and Whiting (1947), Ryder (1948), Locke and Bliss (1950), and Lockwood (1950). The method of injecting plastics or wax into porous media has the further advantage that a differentiation will be made between effective (i.e., interconnected) and total pore space: the plastic, of course, will reach only the former.

Instead of light rays, X-rays can be used for investigating the pore space of porous media. However, such methods are used to obtain the pore size distribution rather than the straight porosity of a porous medium. They will be discussed more fully in section 1.4.

C. *Density methods.* If the density ρ_G of the material making up the porous medium is known, then the bulk density ρ_B of the latter is related to the fractional porosity P as follows:

$$P = 1 - \frac{\rho_B}{\rho_G} \qquad (1.1.2.1)$$

The bulk density can be obtained in several ways. If measuring the outside dimensions and weighing the piece of material (Babcock, 1945) do not give results accurate enough, a volumetric displacement method can be applied. For fluid to be displaced, a non-wetting fluid such as mercury has to be used which (it is hoped) will not enter the porous medium—as long as the pore radii are not too large (Seil, Tucker and Heiligman, 1940). The displacement method can be improved upon if the porous medium is coated with a suitable material before immersion (Melcher, 1921), in which case corrections have to be made to allow for the coating. Weighing under mercury is another method of obtaining bulk density (Westman, 1926; Athy, 1930; King and Wilkins, 1944). In this method, a heavy mass usually has to be attached to the sample in order to submerge it.

The density of the material of which the porous medium is composed can be obtained by crushing the latter and weighing the parts; determining their volume by a displacement method thereupon yields effective porosity. An alternative possibility of porosity determination is provided by measuring the change in weight of a soaked porous medium outside the liquid and by measuring directly the volume of liquid that was soaked up. The soaking method is quite an old one; it has been applied in recent years by Gorodetskiĭ (1940), Schumann (1944), Lepingle (1945), Saunders and Tress (1945), Cross and Young (1948), Eichler (1950), Lakin (1951), and Pollard and Reichertz (1952).

D. *Gas expansion method.* The basic principle of the gas expansion method is direct measurement of the volume of air or gas contained in the pore space. This can be achieved either by continuously evacuating the air out of the specimen, or, in the more modern versions of the method, by enclosing the specimen of known bulk volume and a certain amount of gas (or air) in a container of known volume under pressure, then connecting this with an evacuated container of known volume. The new pressure of the system is read off and permits one to calculate immediately from the gas laws (Boyle-Mariotte) the volume of gas that was originally in the porous medium. It is obvious that this method gives effective porosities.

The gas expansion method was originally employed by Washburn and Bunting (1922). Equipment based on the gas-expansion principles has been refined and described by McGee (1926), Stevens (1939), Page (1947), Hofsass (1948), Kaye and Freeman (1948), Rall and Taliaferro (1949), and Beeson (1950).

Following Beeson (1950), a typical apparatus for porosimetry by the gas expansion method and its operation may be described as follows. As shown in figure 1, the apparatus consists of a gauge A connected to the sample chamber H through the needle valve K. Similar valves J and L are in the line to control the flow of a gas, such as helium, into the gauge and to exhaust the gas from the sample chamber. This is equivalent to two inter-connected chambers, with the

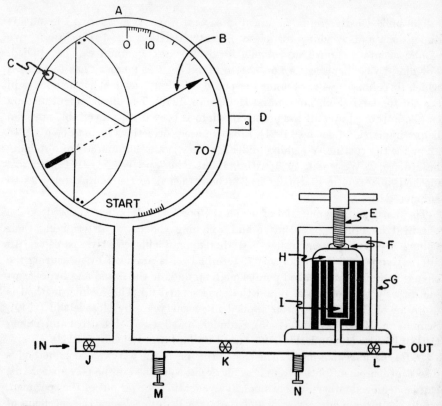

Figure 1. A typical apparatus for the measurement of porosity by the gas-expansion method. (After Beeson, 1950.)

gauge and the connecting lines being one, and the sample chamber and connecting lines being the other. Valves M and N have been added to permit adjustment of the zero point by altering the amount of either valve stem included in the system.

In operation, helium is forced into the gauge chamber until the needle B on the dial attains the "start" position (100.00 psi), as seen through the illuminated eyepiece C mounted on a swivel attached to the gauge. This reading of the gauge, as well as all others, is made with the attached buzzer D in operation, to eliminate the effect of friction on the position of the needle. The gas is then expanded into the sample chamber and the position of the needle is noted. This reading, together with a calibration table, yields the volume of the sample contained in the chamber.

The calibration table is obtained by performing the above procedure with samples of known volume, such as chrome-alloy ball-bearings that have been measured accurately with a micrometer. This method of calibration minimizes errors due to deviations of the gas from the perfect gas law.

The gas expansion method is probably most generally used at present. It is fast, relatively accurate, and leaves the sample in an undisturbed state so that other tests can be performed immediately.

E. Other methods. Methods other than those outlined above have been proposed for porosity measurements. As stated in the beginning of this section (1.2.1), one has to determine, for a porosity measurement, both bulk volume and pore (or solid) volume of a piece of material. The "other" methods are concerned with rather unorthodox measurements of the latter.

One such method is the adsorption of liquids to the internal surface; the surface area can then be determined, and, if the pore geometry is known, the porosity. These methods will be outlined in detail in section (3.2) on adsorption of fluids by porous media.

Another way to obtain the pore volume is to measure the resistance of the medium to flow through the pores. According to some formulas, to be developed in the chapters on fluid flow, the pore volume can be obtained if certain assumptions about pore geometry are made (cf. the Kozeny theory, sec. 6.3.2). Although this method is in general not well founded, Drotschmann (1943) claims that it is working well when applied to plates in batteries.

1.2.2. Specific Cases. The methods of porosity measurement have been applied and adapted to many cases. Lists of the range of porosity values for such cases have been obtained.

In table I, we supply a list of representative values of the porosity of a variety of substances which have been reported in the literature. The values listed under "range" should not be regarded as *the* maximal and minimal values possible, but rather as values indicating the range in which the porosity of the respective substances is likely to be found. In the column headed "Literature," we give references to the original papers from which the values shown in the table were taken.

1.3. The Measurement of Specific Surface

1.3.1. Methods. As outlined in section 1.1, the specific surface is defined as the surface of the pores per unit bulk volume of the porous medium. It is therefore expressed as a reciprocal length.

There are several methods of measuring it. The following is a compilation of the more common methods from the literature.

TABLE I
Representative Values of Porosity for Various Substances

Substance	Porosity range (Porosity in %)	Literature reference
Berl saddles	68–83	Carman, 1938
Raschig rings	56–65	Ballard and Piret, 1950
Wire crimps	68–76	Carman, 1938
Black slate powder	57–66	Carman, 1938
Silica powder	37–49	Carman, 1938
Silica grains (grains only)	65.4	Shapiro and Kolthoff, 1948
Catalyst (Fischer-Tropsch, granules only)	44.8	Brötz and Spengler, 1950
Spherical packings, well shaken	36–43	Bernard and Wilhelm, 1950
Sand	37–50	Carman, 1938
Granular crushed rock	44–45	Bernard and Wilhelm, 1950
Soil	43–54	Peerlkamp, 1948
Sandstone ("oil sand")	8–38	Muskat, 1937
Limestone, dolomite	4–10	Locke and Bliss, 1950
Coal	2–12	Bond et al., 1950
Brick	12–34	Stull and Johnson, 1940
Concrete (ordinary mixes)	2–7	Verbeck, 1947
Leather	56–59	Mitton, 1945
Fibre glass	88–93	Wiggins et al., 1939

A. Optical methods. In analogy to the optical methods of porosity measurements, the specific surface can also be determined from photomicrographs obtained according to the procedures outlined in section 1.2.1. If it is somehow possible to determine the ratio of the circumference of the pores to that of the total area of the section, then there is a simple relationship with the required specific area. Care must be taken, however, to allow for the over-all magnification of the photograph; since the dimension of the specific area is a reciprocal length, it will come out n times too small if the magnification of the picture was n-fold. The value of the specific area as determined from the picture must therefore be multiplied by the magnification factor employed in order to get the correct specific area of the sample.

The determination of the specific internal area from a photomicrograph is best accomplished by a statistical method due to Chalkley, Cornfield and Park (1949). Let a needle of length r fall a great number of times upon the picture. Count the number of times the two end points of the needle fall into the interior of the pores, denoted by h for "hits," and also count the number of times the needle intersects the perimeter of the pores, denoted by c for "cuts." Then, in a very large number of throws, it is a mathematical result that

$$\frac{rh}{c} = 4 \frac{\text{volume of pores}}{\text{surface of pores}} \qquad (1.3.1.1)$$

if the magnification of the picture is 1. The bulk volume of the specimen is $1/P$ times the volume of the pores, therefore one obtains

$$S = \frac{4Pc}{rh} \qquad (1.3.1.2)$$

and, allowing for the n-fold magnification of the photomicrograph, this yields

$$S = \frac{4Pcn}{rh} \qquad (1.3.1.3)$$

B. *Methods based on adsorption.* The adsorption of a vapour by a solid surface has a connection with the area of that surface. A fairly reliable determination of the internal surface area can be obtained upon this basis. There are several theories of adsorption of fluids by porous media which will be discussed in a separate section (3.2.1) of this study. The application of those methods to internal area determination will also be studied there in detail and the reader is referred to that section (3.2.2).

C. *Methods based on fluid flow.* Formulas have been developed which claim to relate the rate of flow of fluids through porous media to their specific surface area. These formulas are known as the Kozeny equation and the Kozeny-Carman equation. Because of their importance in the theory of flow, they will be discussed in detail later (cf. sec. 6.3.2).

The Kozeny equation has been used extensively for the determination of the specific surface of porous media; the particular applications of that equation to such a purpose, however, will be discussed in connection with the exposition of the theory leading thereto (cf. sec. 6.4.1).

D. *Other methods.* Some other methods of surface measurement have also been proposed.

One can, for instance, use the amount of heat conduction by a gas in a porous medium for this purpose. This heat conduction depends on the speed of self-diffusion within the gas which, in turn, is determined by the pore structure. In this way, a method of area determination is obtained (Kistler, 1942).

A surface meter for unconsolidated porous media can be developed in a very neat way if the latter can be fluidized (Bell, 1944). Such a meter is based upon the assumption that the specific surface (per weight) is a linear function of the reciprocal weight of particles that have to be added to a fluidizing agent (e.g., water) in order to obtain a certain optical opacity of the mixture.

In analogy with surface adsorption, there also exists an ionic adsorption phenomenon by which ions from an electrolytic solution are adsorbed to a solid in contact with it. Obviously, this yields another possibility to determine the specific surface of a porous medium (Schofield and Talibuddin, 1948).

In addition, there are some further methods that can be applied to determine the specific surface of a porous medium. Such methods are based upon the catalytic reaction and the direct chemical reaction of the surface with certain other substances. Similarly, the heat of wetting and certain exchange reactions of radioactive substances can also be used (Zimens, 1944).

1.3.2. Specific Cases. The above-listed methods of area determination in porous media have been applied to many specific cases, and tables of representative values have been obtained.

TABLE II

REPRESENTATIVE VALUES OF SPECIFIC SURFACE FOR VARIOUS SUBSTANCES

Substance	Specific surface range (Specific surface in cm^{-1})	Literature reference
Berl saddles	3.9–7.7	Carman, 1938
Raschig rings	2.8–6.6	Ballard and Piret, 1950
Wire crimps	2.9×10–4.0×10	Carman, 1938
Black slate powder	7.0×10^3–8.9×10^3	Carman, 1938
Silica powder	6.8×10^3–8.9×10^3	Carman, 1938
Catalyst (Fischer-Tropsch, granules only)	5.6×10^5	Brötz and Spengler, 1950
Sand	1.5×10^2–2.2×10^2	Carman, 1938
Leather	1.2×10^4–1.6×10^4	Mitton, 1945
Fibre glass	5.6×10^2–7.7×10^2	Wiggins et al., 1939

We supply, in table II, a list of such representative values for various cases. As previously, "range" is to mean the representative range into which the area values of the substances concerned are likely to fall, and not the utmost extremes possible. Again, we also supply references to the literature whence the values shown were obtained.

1.4. The Determination of Pore Size Distribution

The pore size distribution function $\alpha(\delta)$ was defined in section 1.1. Methods have been proposed in the literature to determine it without, however, such an exact definition of it as has been given in section 1.1. The investigators simply talk about "pore diameters" without specifying in any way what is meant by this term. No evaluations exist, therefore, to determine whether the methods proposed actually do measure $\alpha(\delta)$ as defined in section 1.1 or something else.

The method most frequently used is that of injection of mercury into the pore system. In this manner, a capillary pressure curve is obtained which, in turn, can be interpreted in terms of pore size distribution. A detailed account of the method and references to many applications will be given in connection with the discussion of capillarity (cf. sec. 3.4.3).

Surface adsorption has also been used to obtain indirectly a curve for pore size distribution. In a similar fashion capillary condensation in small spaces has been used for obtaining pore size distribution curves. These methods will be discussed in the section dealing with the respective phenomena (cf. sec. 3.2.2).

If the size and frequency of the maximum pores that permit the passage of a fluid through the porous medium only is required, a method forcing gas bubbles through the porous medium and measuring threshold pressures can be employed (Knöll, 1940).

More direct methods using the principles of X-ray scattering, have been employed (Brusset, 1948; Ritter and Erich, 1948; Shull, Elkin and Roess, 1948; Avgul et al., 1951). Another method is to break down the porous medium by crushing it more and more finely. At each stage of crushing the porosity can be measured, which in turn permits evaluation of the pore size distribution as the larger pores are progressively destroyed (Gilchrist and Taylor, 1951).

Direct optical methods such as the study of micrographs are, of course, also available. Procedures employing such methods were cited in connection with porosity measurements, where they give excellent results. Their application to pore size distribution studies, however, is somewhat problematic. The reason for this is, of course, that the void space shown in a section through the porous medium does not necessarily have much relation to the void space in the medium. It is possible that the pores shown may be all rather large in one direction, but small in the other direction, normal to the section. Thus, at least some statistical considerations should be kept in mind if micrographic methods are to be used for pore studies.

The methods mentioned above yield the cumulative pore size distribution curve $f(\delta)$, from which the differential pore size distribution curve $\alpha(\delta)$ has to be calculated. A typical example is given in figure 2. This particular pore size distribution has been determined by the injection of mercury into a piece of limestone which showed very large (vuggy) and very small pores. The maximum point of $\alpha(\delta)$ has obviously not been obtained; the method breaks down for very small openings.

1.5. The Correlation between Grain Size and Pore Structure

1.5.1. The Measurement of Grain Size.
It has been stated in section 1.1 that the grain size distribution of unconsolidated porous media can serve as a basis for the investigation of pore properties. In principle, there are similar difficulties with the notion of "grain size" as exist with the notion of "pore size." The grains are irregularly shaped and therefore their "size" is not *a priori* defined. However, at least the concept of "largest diameter" of a grain makes sense geometrically.

The grain size or grain size distribution of an unconsolidated porous medium can be determined by a variety of methods. However, as the subject of grain size determination is only incidental to the study of the physical principles of hydromechanics in porous media, only a list of selected references will be given.

General discussions and reviews of particle size measurement have been given by Krumbein and Pettijohn (1938), Heywood (1938, 1947), Milner (1940), Uren (1943), and Hawksley (1951). The last article mentioned is a good general review listing 218 references. A book by Dalla Valle (1948) also contains

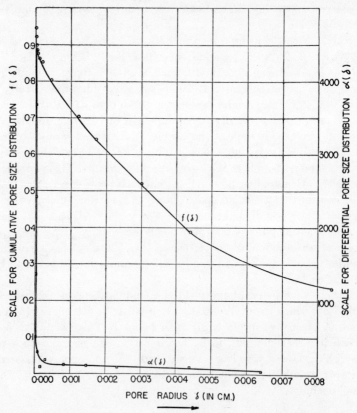

Figure 2. Cumulative pore size distribution curve and differential pore size distribution curve obtained on a piece of limestone by the mercury-injection method. Courtesy of Imperial Oil Limited.

much useful information. In addition to the general references given above, some of the methods applicable to particle size measurement have been described as follows: hydrometric methods (such as sedimentation) by Steinberg (1946), Mills (1948), and Rim (1952); X-ray diffraction methods by DuMond (1947), and Müller (1947); sieving by a multitude of authors, especially Dalla Valle (1948); and the use of the electron microscope for the present purpose by Franklin et al. (1953).

1.5.2. Theory of Packing of Spheres. In order to establish a correlation between grain size and pore size of an unconsolidated porous medium, one has to know something about the packing of the grains as well as about their shape. For even if the grain "size" (i.e., the largest diameter) is known, the shape is still not determined. The grains that pass through a certain sieve-mesh and do not pass through another slightly smaller one, are not necessarily all identical, owing to the irregularity of shape.

Figure 3. Rhombohedral packing of spheres. (After Graton and Fraser, 1935.)

A qualitative visualization of the conditions involved can be obtained by the procedure of assuming "model"-grains, usually spheres, and by studying their modes of packing. Thus, one arrives at a theory of models of porous media which are composed of spheres.

Such models are constructed in the following way. First, the average size of the grains is measured in one way or another and the spheres of the model are envisaged as being of the same diameter. Then the porosity of the original specimen is determined and the packing of the model-spheres is assumed to be arranged in such a fashion that the model has the same porosity as the original.

Therefore in order to be able to construct proper models, one has to investigate the packing of spheres.

The first study of the modes of packing of spheres and the porosity calculated therefrom appears to have been undertaken by Slichter (1899). Since then the theory has been reviewed, refined, and extended by Smith, Foote and Busang (1929), Graton and Fraser (1935), Manegold (1937b), Manegold and Solf (1939), Hrubíšek (1941), Ackerman (1945), Foord (1945), and Busby (1950). Muskat (1937, p. 10) gives a short qualitative review of the theory in his book on flow through porous media, and Leĭbenzon in his monograph (1947, pp. 18–24) gives a concise modern exposition of the theory of Slichter, including the mathematical deductions involved.

The above-mentioned article of Hrubíšek (1941) contains probably the most comprehensive investigation of the geometry of assemblies of spheres to date. It is based upon Minkowski's theory of the geometry of numbers and gives a very complete survey of the applications of this theory to the present problem. The most difficult part is, of course, an enumeration of the possible patterns of assemblies of spheres and mathematical proofs for maximal and minimal properties thereof. Of practical importance are "stable" packings only, i.e., packings where one sphere touches at least four others in such a manner that a least four of the contact-points are not contained in the same hemisphere.

Without going into the mathematics of the theory, for which the reader is referred to the article of Hrubíšek (1941), one can state the principal results as follows:

(a) For any one mode of packing, the porosity of the bed is independent of the size of the spheres.

(b) The porosity of stable beds varies from 0.259 upwards. The system with porosity = 0.259 represents "rhombohedral" (fig. 3) packing. The "thinnest" stable packing of spheres that is now known has been described by Heesch and Laves (cf. Hrubíšek, 1941); it has a porosity of 0.875.

(c) A quantity, n, can be defined which gives the ratio of the area left void by the spheres in a plane cross-section through the centres of adjacent spheres to the area of that cross-section. It is a measure of the pore size of the array. For the different packings, it varies from 0.0931 upwards (cf. Leĭbenzon, 1947, p. 23). It is again independent of the size of the spheres.

It is thus seen that a complete enumeration of all the stable packings of identical spheres has not yet been achieved. The stable packing with minimum porosity is rhombohedral as mentioned above; it is illustrated in figure 3. A stable packing with maximum porosity has not yet been found; it is doubtful if one exists since larger and larger arches of spheres that are stable can be constructed and it is thus possible that no packing with maximum porosity will be found. However, a mathematical proof of the non-existence of a stable packing with maximum porosity has not been achieved either.

1.5.3. Packing of Natural Materials. Natural materials are composed of grains whose shape may deviate appreciably from that of spheres. Moreover, it will often be found that the grains are somewhat cemented together without being fully consolidated, but it would still be desirable to apply some correlations between "grain size" and pore size distribution and other characteristics of the porous medium, such as specific area. Furthermore, the size of the grains will seldom be very uniform. Non-uniformity in size will, in general, permit the smaller particles to fill the spaces between the larger ones and thus appreciably reduce porosity. Contrariwise, angularity of the particles permits bridging with a resulting increase of porosity.

It is only reasonable, therefore, to expect that spherical models as discussed above will not be entirely adequate to represent even the geometrical properties of porous media (not to speak of hydraulic ones). They will help us to understand some of the features better, but actual correlations between grain sizes and pore sizes, such as have been proposed, are based on experimental investigations rather than on theory. A true understanding of the effects, of course, cannot be obtained in this manner either.

Theoretical and experimental studies of the effect of grain-size on pore-size have been conducted by Tickell, Mechem and McCurdy (1933), Nissan (1938), Cloud (1941), Hrubíšek (1941), Rosenfeld (1949), and Griffiths (1952). Kiesskalt and Matz (1951) discussed the dependence of specific area of the grain size distribution.

These studies bear out some general facts such as those mentioned at the beginning of this section (1.5.3); but the correlations obtained do not seem to be generally applicable without the introduction of sufficient "cementation," "compaction" and other "factors"—which, actually, are only undetermined factors to allow adjustment of the actual findings to those one would like to expect. It should be noted that any correlation can be made to fit any set of findings if a sufficient number of indeterminate factors is used.

Thus the correlations developed are at best valid for one type of material or rock strata. To expect that Raschig rings and spherical pebbles, which happen to have been screened out by the same set of sieves, should have identical geometrical pore structures is somewhat over-optimistic.

It appears, therefore, that it is not really a satisfactory procedure to try to represent porous media by models composed of such well-ordered things as identical spheres in regular packings. The character of porous media is disorder rather than order, and one ought to take this into account in attempting any model representations. The use of geometrical models of porous media is therein analogous to the use of dynamical models to describe flow through porous media. In the case of flow, principles pertaining to the general statistics of disordered phenomena have been applied, but in the case of geometrical properties, this has not yet been done. It is to be expected that the proper characterization of geometrical properties of porous media can also be achieved only by the use of statistical methods.

1.6. Rheological Properties of Porous Media

Porous media have been treated so far as rigid, an assumption which cannot always be expected to be true. The geometric quantities describing porous media, as introduced, may themselves be functions of certain dynamic quantities, notably of the prevailing stresses.

The description of stresses in porous media meets with certain difficulties. It is obviously as impractical to give the actual stress-tensor microscopically at every point of the medium, as it is impractical to give the analytical equation of the surface constituting the pore-system.

Thus, some heuristic theory has to be developed which describes the stresses in porous media in macroscopic terms. Gassmann (1951) has developed a rather complete theory of such stresses, but for our purpose, a more elementary theory proposed by Terzaghi (1951) is entirely sufficient. Without going into the details of Terzaghi's theory, one may state that the stresses in a porous medium are two-fold—one part, termed the *neutral stress*, is the stress in the fluid; the second part, called the *effective stress*, is the difference between the total stress prevailing in the fluid-filled porous medium and the neutral stress. It is this stress which produces the deformation of the porous medium.

The macroscopic stress in the fluid is hydrostatic, and can therefore be denoted by a scalar p. It is the pressure of the pore-fluid. The total macroscopic stress of the porous medium (including the pore-fluid) may be denoted by a tensor $\mathscr{T}$ with components $\mathscr{T}_{ik}$. In most cases, however, the hydrostatic component of this stress will be the prevailing one, as it is equal to the "overburden-pressure" of the porous medium. We shall denote this scalar overburden pressure by $\mathscr{T}$, too, as it is quite clear from the context whether a tensor or a scalar is implied.

In order to determine the dependence of the various geometrical quantities characterizing porous media upon stress, it is necessary to develop some connection between the stresses and the corresponding strains of both the fluid and the solid. The fluids will be discussed at length in Part II of this book, and, as for porous media, the following remarks can be made.

The simplest connection between geometrical quantities referring to the porous medium and the stresses will be obtained by assuming the porous medium as elastic, in which case Hooke's law obtains. The effect of this will be that all the geometrical quantities relating to the pores are linear functions of the effective stress. By the boundary conditions, however, the fluid pressure p and the total stress $\mathscr{T}$ will generally be prescribed so that it appears that for a fixed total stress, the pore geometry is a linear function of p. With reference to the porosity P, this leads to the following equation:

$$dP/dp = \beta_m(\mathscr{T}) \qquad (1.6.1)$$

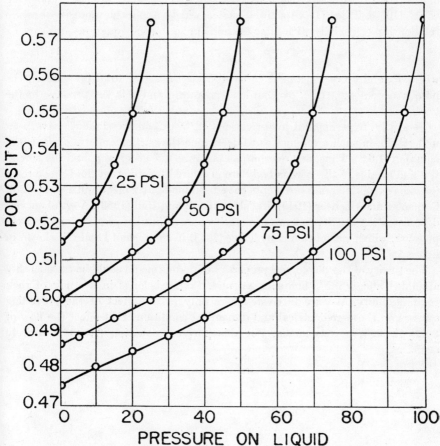

Figure 4. A set of curves demonstrating the dependence of porosity on overburden pressure and on fluid pressure, in the case of kaolin. (After Tiller, 1953.)

In a more complicated case, the rheological equation of the porous medium may be assumed as different from that of Hooke. The medium may be plastic, in fact it may itself behave externally almost like a "fluid" (cf. Ferrandon, 1950). Clay, for instance, will consolidate if any fluids contained in it are withdrawn. This will show up by the fact that its porosity as well as the other geometrical quantities characterizing it, are not linear functions of p—they are not even reversible. Nevertheless, it will be possible to represent the porosity as some function of p, with the components of the total stress ($\mathscr{T}$) as parameters:

$$P = P_{\mathscr{T}}(p) \tag{1.6.2}$$

and similarly with the other geometrical quantities of interest. Following Grace (1953), Tiller (1953) showed that a widely applicable porosity-pressure relationship (especially with filter cakes) is given by the equation

$$P = P_0(\mathscr{T} - p)^{-c} \tag{1.6.3}$$

where P_0 and c are constants; as usual $\mathscr{T}$ is the scalar total pressure on the porous system (i.e., the "overburden" pressure), and p is the pressure in the fluid.

Generally, experimental investigations of "rock compressibility" have been undertaken in such a manner as to determine the particular form of $P_{\mathscr{T}}(p)$ in equation (1.6.2). Thus, an "overburden pressure" ($\mathscr{T}$) is chosen, and the porosities and strains of the porous body are measured for different fluid pressures p (or *vice versa*). Such measurements have been made by Botset and Reed (1935), Carpenter and Spencer (1940), Bal'shin (1949), and Hall (1953). A set of curves, demonstrating the dependence of porosity on overburden pressure and on fluid pressure, which was obtained by Tiller (1953) in the case of kaolin, is shown in figure 4.

The theory of the rheological properties of porous media could be considerably amplified and refined, but such an undertaking belongs in a study of rock compressibility. We are interested here only in the effect of the prevailing stresses on those geometrical and dynamical conditions that effect the flow of fluids through porous media. For this purpose, equation (1.6.2) is entirely sufficient.

PART II

FLUIDS

2.1. General Remarks

The subject of hydromechanics in porous media is concerned with the pore space of a porous medium as filled with some fluids. It is therefore also necessary to investigate the mechanics of fluids that may fill the pore spaces, i.e., of liquids and of gases.

There are many good reference books on the mechanics of liquids and gases, and thus the present chapter is but a very incomplete review of the subject for the convenience of the reader. Only those aspects of the theory will be stressed which will be of importance later.

From the outset, there are two possible aspects of the mechanics of fluids: a macroscopic and a microscopic. The macroscopic aspect is borne out by what one might call the "continuous matter theory" of such fluids, meaning that the fluid is treated as a continuous medium, its motion being determined if the motion of every material point of the fluid is given by mathematical equations. The microscopic aspect is obtained if the molecular structure of the fluid is taken into account. Usually the microscopic aspect will effect but minor corrections to the equations of continuous matter theory.

Of particular importance with respect to the flow through porous media is the interaction of fluids with surfaces, and the final sections of this part will therefore be devoted to that subject.

2.2. Continuous Matter Theory

2.2.1. The Fundamental Equations.
The motion of a fluid, if the latter be regarded as a continuum, is geometrically described if the position of every material point of the fluid is known at every time-instant. There are three kinds of physical conditions which determine such motion. The first is the continuity condition, the second the rheological equation of state, the third Newton's law of motion.

These physical conditions are expressed mathematically as a system of differential equations. An additional set of initial or boundary conditions is needed to make the problem fully determined. Such boundary conditions, for example, specify whether or not the particular fluid is sticking to the walls of a container. Different combinations of rheological equations and boundary conditions will determine whether the fluid is a liquid or a gas, whether it is viscous or non-viscous, etc.

The *continuity condition* and *Newton's law* of motion are well known; for example, Lamb (1932) expressed them in a representation suitable to the description of a continuous medium.

The *rheological condition* is the connection between the stresses and the strains in the fluid (and their time derivatives). In the case of an "ideal" fluid it is assumed that there are no shearing-stresses possible and that the fluid is incompressible. In general, however, we shall assume that the fluids are viscous and compressible. The conditions applied to account for viscosity (introducing a constant, μ, termed "viscosity") are well known (see, for example, Lamb, 1932), and for the compressibility we shall assume the following general equation (Muskat, 1937, p. 131):

$$\rho = \rho_0 \left(\frac{p}{p_0}\right)^m e^{\beta_f(p - p_0)} \qquad (2.2.1.1)$$

where ρ is the density and p the hydrostatic pressure. The quantities m and β_f are constants. The particular fluids of significance may now be classified as follows (after Muskat, *loc. cit.*):

Liquids: $m = 0$

    Incompressible liquids $\beta_f = 0$

    Compressible liquids $\beta_f \neq 0$

Gases: $\beta_f = 0$

    Isothermal process: $m = 1$

    Adiabatic process: $m = \dfrac{C_V}{C_p}$

where C_V denotes the specific heat at constant volume and C_p the specific heat at constant pressure.

Finally the *initial* and *boundary conditions* will determine, first, the shape and walls of the container of the fluid, second, the external conditions (such as the throughput or the pressure drop) and, third, the interaction between the fluid and the walls. If the fluid is viscous, it is generally assumed that it sticks to the walls.

2.2.2. Special Cases of Viscous Fluid Flow. *A. The Navier-Stokes Equation.* The set of conditions outlined in section 2.2.1 can be combined to form various differential equations which are applicable to different kinds of fluids. The best known of these equations is that of Navier and Stokes (see Lamb, 1932). It is applicable to incompressible viscous fluids. Because of its fundamental importance it is re-stated here:

$$\boldsymbol{v} \operatorname{grad} \boldsymbol{v} + \partial \boldsymbol{v}/\partial t = \boldsymbol{F} - (1/\rho) \operatorname{grad} p - (\mu/\rho) \operatorname{curl} \operatorname{curl} \boldsymbol{v} \qquad (2.2.2.1)$$

Here $\boldsymbol{v}$ is the local velocity-vector of a point of the fluid, t is time, $\boldsymbol{F}$ the volume-force per unit mass, and, as before, p, μ and ρ are respectively pressure, viscosity, and density of the fluid. The boundary conditions prescribe that $\boldsymbol{v} = 0$ at the walls of the container.

The structure of the Navier-Stokes equation and the boundary conditions make the analytical solution of (2.2.2.1) very difficult. One is therefore led to look for some approximation of (2.2.2.1) which could be treated analytically with more ease. It may be observed that the Navier-Stokes equation is considerably simplified if μ is set equal to zero—in fact, one then obtains the well-known equation of Euler for non-viscous flow where solutions may be obtained from a flow-potential. However, the boundary condition ($\boldsymbol{v} = 0$ at the walls) does not contain the viscosity and thus remains unaffected by the assumption of $\mu = 0$. One can, therefore, state that apart from a thin boundary layer at the walls, a slightly viscous fluid behaves like a non-viscous one. By introducing a sufficient amount of epsilontics, the order of magnitude of the boundary layer can be estimated in terms of the order of magnitude of the viscosity, and an approximate equation (which is analytically relatively simple) for the viscous fluid inside the boundary layer can be deduced from equation (2.2.2.1). One thus arrives at a "boundary layer theory." The latter was originated by Prandtl. The boundary layer has actually been made visible experimentally by ingenious experiments (Koncar-Djurdevic, 1953).

B. The Hagen-Poiseuille Equation. The Navier-Stokes equation can be solved *exactly* for a *straight circular tube*. In this case, the term $\boldsymbol{v}$ grad $\boldsymbol{v}$ is zero because of the orthogonality of $\boldsymbol{v}$ and the gradient of all the components of $\boldsymbol{v}$. The rest of the equation can be integrated quite easily for the prevailing boundary conditions. In terms of total volume throughout Q through a circular tube of radius a and length h, the pressure drop from end to end being Δp, the solution, which is called "Hagen-Poiseuille equation," becomes (Lamb, 1932):

$$Q = \left(\frac{\pi}{8}\right)\left(\frac{\Delta p}{h}\right)\frac{a^4}{\mu}. \tag{2.2.2.2}$$

C. Turbulence and Reynolds Number. Theory and experiment show that for high flow velocities the flow pattern becomes transient although the boundary conditions remain steady: eddies are formed which proceed into the fluid at intervals. For any one system, there seems to be a "transition point" below which steady flow is stable. Above the "transition point" the steady flow is more and more likely to become unsteady and to form eddies upon the slightest disturbance. The steady flow is often termed *laminar*, the flow containing eddies *turbulent*. In turbulent flow, the law of Hagen-Poiseuille is no longer valid—the resistance (pressure drop) becomes a quadratic function of the throughput.

Although the transition point has been calculated from the Navier-Stokes equation for certain simple systems, it is obvious that such a calculation is a very difficult undertaking. One has therefore to recur to experiments to determine when turbulence will set in. If some systems can be shown to be dynamically similar, then the transition point in one system will have a corresponding point in the dynamically similar system which can then be calculated.

It has been shown by Reynolds that circular tubes are dynamically similar, as far as the Hagen-Poiseuille equation is concerned, if the following "Reynolds number" (denoted by Re) is the same:

$$Re = 2\rho a v/\mu \qquad (2.2.2.3)$$

where all the constants have the same meaning as before, and v is now the average flow velocity in the tube.

It must be expected, therefore, that turbulence will occur in any straight tube if a certain Reynolds number is reached. This critical Reynolds number has been determined to be in the neighbourhood of 2200 (Joos, 1947, p. 204).

Formula (2.2.2.3) contains only the radius a of the tube, apart from constants referring to the fluid. It has, therefore, been indiscriminately applied to *any* tube, straight or not, with rather devastating results, for the calculation of flow through a straight tube is based on the neglect of the term **v** grad **v** in the Navier-Stokes equation. The Reynolds number is, therefore, a valid criterion of dynamical similarity *only* if **v** grad **v** vanishes in the two systems to be compared. This term, however, is anything but zero in a curved tube.

This statement has been confirmed experimentally by Comolet (1949), who has shown that the critical Reynolds number at which water flowing in a tube becomes turbulent is lowered greatly by a slight curvature of the tube. This was proven by making a quantitative study using a flexible tube projecting from a reservoir. Recently Topakoglu (1951) made a theoretical approach to the problem and showed explicitly that the flux is a function of both the Reynolds number *and the curvature of the tube*.

We should therefore like to emphasize once more that identical Reynolds numbers are *not* a sufficient condition to ensure dynamical similarity between two systems consisting of flow channels; in this instance the Reynolds number has been greatly abused, as *it is also necessary that the channels be straight*.

2.3. Molecular Flow

The theorems obtained from the continuous matter theory have been observed experimentally to be in need of correction if the distance between the walls confining the fluid is of the same order of magnitude as the free molecular path length in the fluid. It has been found that the essential correction needed is at the boundary condition which stated that $v = 0$ at the walls. In fact, it has been observed that the molecular nature of the fluid essentially effects a "slip" at the walls such that $v \neq 0$ at the (fixed) walls. It is to the credit of Kundt and

Warburg (1875) that they observed this fact first. As a consequence, the quantity of gas flowing through a capillary is larger than would be expected from Poiseuille's formula. Warburg (see Klose, 1931) applied this concept to obtain a "slip correction" of Poiseuille's formula by adding a constant term to the latter. This means that under a zero pressure differential there is a finite "slip flow" through a capillary. At other pressure differentials, the flow is always bigger by this finite value than would be calculated from Poiseuille's formula.

Experimental investigations prompted Knudsen (1909) to propose yet another condition for the total volume rate of gas flow Q (measured at pressure p) through a capillary of radius a and length h. He postulated:

$$Q = \frac{4}{3} \sqrt{\frac{2\pi R \mathfrak{T}}{M}} \frac{a^3}{h} \frac{\Delta p}{p}. \tag{2.3.1}$$

Here, R is the gas constant, $\mathfrak{T}$ the (absolute) temperature and M the molecular weight of the gas. According to the kinetic theories of gases, the mean free path length λ, is given by

$$\lambda = c \frac{\mu}{\bar{p}} \sqrt{\frac{R \mathfrak{T}}{M}} \tag{2.3.2}$$

where c is a constant approximately equal to 2, μ the viscosity, and $\bar{p}$ the mean pressure. Thus Knudsen's equation can be rewritten as follows:

$$Q = \frac{4}{3} \sqrt{2\pi} \frac{\lambda}{c\mu} \frac{a^3}{h} \bar{p} \frac{\Delta p}{p} \tag{2.3.3}$$

where λ is to be measured at the mean pressure.

It soon became established that Knudsen's equation describes the flow correctly if the mean free path is very large compared with the tube radius a (see, for example, Klose, 1931), but that Poiseuille's equation has to be used if the mean free path is very small. It is therefore a ready conjecture that, for intermediate cases, the two equations have to be combined. This has been proposed, for example, by Adzumi (1937) who thus arrived at the following equation:

$$Q = \frac{\pi a^4 \bar{p} \Delta p}{8 \mu h p} + \varepsilon \frac{4}{3} \sqrt{2\pi \frac{R \mathfrak{T}}{M}} \frac{a^3}{h} \frac{\Delta p}{p}. \tag{2.3.4}$$

Here, ε is a dimensionless proportionality factor which has generally values of about 0.9 in the case of single gases, and values of about 0.66 in the case of gaseous mixtures. It may thus be assumed safely that ε has some constant value for any flow phenomenon of interest. It is called "Adzumi constant."

Later on, the Adzumi equation was modified by Deryagin, Fridlyand and Krȳlova (1948) and by Arnell (1946) with the intention of applying it to model-porous media. These modifications will be discussed in detail later, in connection with the models of porous media for which they were intended.

2.4. The Interaction of Fluids with Surfaces

2.4.1. Adsorption. The boundary conditions at the walls of a container of a fluid are not always as simple as outlined in either sections 2.2 or 2.3. Especially if the fluid is a vapour near its condensation point, further remarkable effects may occur.

Thus, the molecules of the fluid may be *adsorbed* by the walls of the container. This signifies that within a few molecular distances from the wall there is a strong attractive potential between it and the molecules of the fluid. Thus, a much greater concentration of molecules of the fluid may be found at the walls of the container than in its interior. The general aspects of adsorption may be found in any text-book on colloid chemistry (e.g., Hartman, 1947) or in the fundamental monograph of Brunauer's (1943).

During the process of adsorption, energy is liberated at the walls of the container. This energy is called heat of adsorption. Thus if any pressure (or other) measurements are made during the process of adsorption, the temperature of the system has to be carefully watched. In general, conditions are arranged in such a manner that the liberated heat is conducted away so that the process is isothermic. One therefore refers to curves obtained in such a manner as adsorption *isotherms*. In particular, it is understood that an "adsorption isotherm" is a plot of (molecular) pressure (in the centre of the container) *versus* the number of molecules of the fluid on the walls of the container, and care must be taken that the process of adsorption is isothermic. Examples of some isotherms are given in figure 5 (after Emmett, 1948).

Because of the large potential present in the adsorbed layers, the rheological state of the fluid inside the container may be different from that on the walls. In particular, if the fluid is a vapour near its condensation point, the adsorbed molecules may well form a film of *liquid* along the walls.

The quantitative aspects of adsorption will be dealt with in Part III on "Hydrostatics in Porous Media," since this is practically the only instance where large surfaces come into contact with fluids in such a manner that adsorption may become significant.

2.4.2. Interfacial Tension. If a liquid is in contact with another substance (gas or solid), there is free interfacial energy present between the two (see, for example, Adam, 1941). This means that a certain amount of work has to be performed in order to separate a liquid from, say, a solid. The interfacial energy arises from the inward attraction of the molecules in the interior of a substance upon those at the surface. Since a surface possessing free energy contracts if it can do so, the free interfacial energy manifests itself as *interfacial tension*. The interfacial energy between a substance and the vacuum is called *surface* *tension*. Surface tensions will be denoted in this monograph by γ, with subscripts S, L, A, etc., for solid, liquid, air, etc., respectively. Correspondingly, interfacial tension will be denoted by the same symbol γ, but with two indices indicating to which substances it refers. It should be noted, however, that the

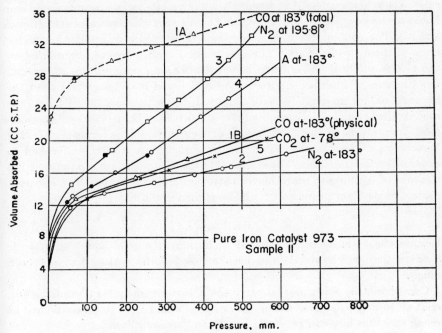

Figure 5. Examples of adsorption isotherms on iron catalyst for various gases near their boiling points. (After Emmett, 1948.)

terms "surface tension" and "interfacial tension" are often used indiscriminately.

The work W_{SL} required to separate a unit area of, for example, a solid from a liquid is related to the corresponding surface and interfacial tensions by an equation established by Dupré as follows:

$$W_{SL} = \gamma_S + \gamma_L - \gamma_{SL}. \qquad (2.4.2.1)$$

If we assume now that two fluid phases (for instance, liquid and air) are at one point in contact with a solid, then the surface tensions between the various phases can be calculated from (2.4.2.1). It can be shown that equilibrium is possible only if the interface between the air and the liquid forms a definite angle with the solid. This angle has been termed "contact angle" and will be denoted here by θ. It is determined by the condition

$$\cos \theta = (\gamma_{SA} - \gamma_{SL})/\gamma_{LA}. \qquad (2.4.2.2)$$

Equation (2.4.2.2) shows that no equilibrium is possible if $\gamma_{SA} - \gamma_{SL}$ is larger than γ_{LA}. In this case, the liquid will spread indefinitely over the solid.

According to equation (2.4.2.2), the cosine of the contact angle is given by the ratio of the energy released in forming a unit area of interface between solid and liquid to the energy expended in forming a unit area of interface between liquid and air. The contact angle, thus, depends on a proper determination of the areas of these two interfaces. The area of the interface between liquid and air can always safely be assumed to be equal to the apparent surface of contact, but the area of interface between solid and liquid may be greater than the apparent or "macroscopic" area of contact. If this is the case, the surface of the solid is called "rough." It depends, of course, solely upon the accuracy of measurement how much of the "roughness" of this surface is incorporated in the "macroscopic" area in the first place. Nevertheless, it is often convenient to express "surface roughness" by stating that the "true" surface is σ (where σ is a measure of roughness) times bigger than the "apparent" area. The apparent contact angle θ' on such a rough surface is then given by

$$\cos \theta' = \sigma(\gamma_{SA} - \gamma_{SL})/\gamma_{LA} = \sigma \cos \theta \qquad (2.4.2.3)$$

because the surface energy released in forming the solid-liquid interface is σ times that which would have been released if the surface had not been rough.

Care should be taken not to carry the above argument *ad absurdum*. For, relating to molecular dimensions, any surface would have to be considered as rough and thus any contact angle is only an "apparent" one.

The theoretical aspects of surface tension and capillarity have been discussed and reviewed by Brown (1947), Shuttleworth and Bailey (1948), Burdon (1949), Gurney (1949), Koenig (1950), and Elton (1951). Prigogine and his co-workers (Prigogine, 1950; Prigogine and Maréchal, 1952; Prigogine and Saroléa, 1952) developed a molecular theory of the surface tension of solutions; MacLellan (1952) advanced a statistical-mechanical theory of the same phenomenon. However, these theories are beyond the scope of this study.

With interfacial tension, there is the complication that it may not be the same if a liquid is either advancing or receding on a solid. One thus observes a possible hysteresis effect in interfacial tension, and so also in contact angles. This phenomenon has been reviewed and discussed by Cassie (1948) and by Shuttleworth and Bailey (1948). For our purposes it is sufficient to realize the possibility of its occurrence.

The quantities γ_{LA}, etc., are specific to the two phases which the indices signify. There is no general expression for the γ's which would be applicable to any fluid or solid. Thus, the contact angle is specific to all *three* phases involved. In particular, if the fluid phases are not changed at all, this angle depends on the condition of the solid. Measurements of interfacial tensions have been made for many combinations of substances and are available in text-books on physical chemistry.

If the properties of the solid are such that it makes the contact angle less than 90° towards one of the fluid phases, one says that the solid has a preferential *wettability* to that phase. The term "wettability" has therefore a relative meaning only; moreover, it is also subject to hysteresis in the same manner as the surface tensions. Thus of two fluids one may preferentially wet a solid if it is receding from the solid, but not if it is advancing.

2.5. Miscible Fluids

Particular phenomena are encountered if two fluid phases, which are completely miscible, are in contact with a bounding surface. In this case the discussion in terms of capillarity and interfacial tension is no longer applicable and new concepts have to be developed.

Miscibility of two fluid phases implies that there cannot be any equilibrium other than that achieved by total mixing of the two phases so that the concentration of the one phase within the other is constant throughout the whole system under consideration. Therefore, no distinct interface between the two phases can exist and all the phenomena of interest are transient.

It turns out that the investigation of the dynamics of miscible fluids is much more difficult than the investigation of immiscible fluids, chiefly because of their fundamentally transient character. The speed of mixing of the two fluids is, on the one hand, conditioned by the speed of the internal molecular diffusion, and, on the other, by the mechanical convection imposed on the mixture. Thus, considering the simplest case, *viz.*, that of a soluble fluid being introduced into a fluid flowing slowly through a small-bore tube, the soluble substance spreads out under the combined action of molecular diffusion and the variation of velocity over the cross-section. Thus, the invading soluble substance at the centre of the tube moves much more rapidly than near the edge. If radial diffusion were absent, an ever-lengthening "needle" of invading fluid would extend down the tube. However, in actuality the fluids interdiffuse radially, "washing out" the needle. An approximate treatment of this simple case has been given by Taylor (1953); Taylor's theory has been reviewed by Von Rosenberg (1956).

Following Taylor (1953), the dispersion of the invading fluid has, thus, to be separated into two contributions originating in convection and in radial diffusion. Considering first convection, we may note that in a circular tube of radius a the velocity v at distance r from the axis is, according to the analysis of Hagen-Poiseuille:

$$v = v_0(1 - r^2/a^2) \qquad (2.5.1)$$

where v_0 is the maximum velocity at the axis. If the solute is distributed symmetrically, the concentration C is:

$$C = f(x, r), \qquad (2.5.2)$$

where $f(x, r)$ denotes "a function of" x and r.

After the elapse of the time t, the concentration will be

$$C = f(x - vt, r).$$

Accordingly, the mean value C_m of the concentration over a cross-section of the tube is defined by

$$C_m = \frac{2}{a^2} \int_0^a Cr\,dr. \qquad (2.5.3)$$

Taylor now considered two cases. First, assuming that the space between two planes $x = 0$ and $x = X$ (X/a being small) is filled initially with solute of concentration C_0, it is seen that the amount which lies between r and $r + \delta r$ is constant during the flow and equal to $2\pi r C_0 X \delta r$. The solute will be distorted into a paraboloid defined by the following equation:

$$x = v_0 t(1 - r^2/a^2). \qquad (2.5.4)$$

The total amount of solute between x and $x + \delta x$ is therefore given by $2\pi r C_0 X (dr/dx)\delta x$, and, inserting dr/dx from equation (2.5.4), one obtains

$$C_m = C_0 X/(v_0 t). \qquad (2.5.5)$$

The mean concentration C_m therefore has the constant value $C_0 X/(v_0 t)$ in the interval $0 < x < v_0 t$ and is zero when $x < 0$ and when $x > v_0 t$.

Second, if solute of constant concentration enters a tube which at time $t = 0$ contains only solvent, one has

$$\left.\begin{array}{l} C = C_0,\ x < 0 \\ C = 0,\ x > 0 \end{array}\right\} \text{ at time } t = 0. \qquad (2.5.6)$$

This case can be treated by imagining that the constant initial concentration for $x < 0$ consists of a number of thin sections of the type considered above, leading to equation (2.5.5). In this manner one finds:

$$\left.\begin{array}{l} C_m = C_0,\ x < 0 \\ C_m = C_0(1 - x/(v_0 t)),\ 0 < x < vt \\ C_m = 0,\ x > v_0 t \end{array}\right\} \qquad (2.5.7)$$

After considering dispersion by convection alone, Taylor considered the effect of molecular diffusion. It is natural to assume that the concentration C is

§2.5   MISCIBLE FLUIDS

symmetrical about the axis of the tube so that C is a function of r, x, and t only. The equation of diffusion is then

$$D\left(\frac{\partial^2 C}{\partial r^2} + \frac{1}{r}\frac{\partial C}{\partial r} + \frac{\partial^2 C}{\partial x^2}\right) = \frac{\partial C}{\partial t} + v_0\left(1 - \frac{r^2}{a^2}\right)\frac{\partial C}{\partial x}. \quad (2.5.8)$$

Here D, the coefficient of molecular diffusion, is assumed to be, in approximation, independent of C.

In all the cases which Taylor considered, $\partial^2 C/\partial x^2$ is much less than $\partial^2 C/\partial r^2 + \partial C/\partial r/r$, so that, writing

$$z = r/a, \quad (2.5.9)$$

the diffusivity equation (2.5.8) becomes:

$$\frac{\partial^2 C}{\partial z^2} + \frac{1}{z}\frac{\partial C}{\partial z} = \frac{a^2}{D}\frac{\partial C}{\partial t} + \frac{a^2 v_0}{D}(1 - z^2)\frac{\partial C}{\partial x}. \quad (2.5.10)$$

The boundary condition, which indicates that the walls of the tube are impermeable, is:

$$\partial C/\partial z = 0 \quad \text{at} \quad z = 1. \quad (2.5.11)$$

The solution of the diffusivity equation, even in the simplified form (2.5.10) under the boundary condition (2.5.11), is difficult to achieve; however, Taylor showed that it is relatively easy to obtain an approximate solution under the assumption that the time necessary for appreciable effects to appear, owing to convective transport, is long compared with the "time of decay" during which radial variations of concentration are reduced to a fraction of their initial value through the action of molecular diffusion. This assumption is applicable in "slow" flow.

Since molecular diffusion in the longitudinal direction has been neglected, the whole of the longitudinal transfer of concentration is due to convection. Considering the convection across a plane which moves at constant speed $\tfrac{1}{2}v_0$, and writing

$$x_1 = x - \tfrac{1}{2}v_0 t \quad (2.5.12)$$

(note that $\tfrac{1}{2}v_0$ is the mean speed of flow), equation (2.5.10) becomes

$$\frac{\partial^2 C}{\partial z^2} + \frac{1}{z}\frac{\partial C}{\partial z} = \frac{a^2}{D}\frac{\partial C}{\partial t} + \frac{a^2 v_0}{D}(\tfrac{1}{2} - z^2)\frac{\partial C}{\partial x_1}. \quad (2.5.13)$$

Since the mean velocity across planes for which x_1 is constant is zero, the transfer of C across such planes depends only on the radial variation of C. Under the above assumptions, this radial variation is small and can therefore be calculated from the following equation

$$\frac{\partial^2 C}{\partial z^2} + \frac{1}{z}\frac{\partial C}{\partial z} = \frac{a^2 v_0}{D}(\tfrac{1}{2} - z^2)\frac{\partial C}{\partial x_1} \quad (2.5.14)$$

wherein $\partial C/\partial x_1$ may be taken as independent of z.

A solution of (2.5.14) satisfying the boundary condition is

$$C = C_{x_1} + A(z^2 - \tfrac{1}{2}z^4) \tag{2.5.15}$$

where C_{x_1} is the value of C at $z = 0$ and A is a constant.

Substituting (2.5.15) into (2.5.14), Taylor found

$$A = \frac{a^2 v_0}{8D} \frac{\partial C}{\partial x_1}. \tag{2.5.16}$$

The rate of transfer Q of C across the section at x_1 is

$$Q = -2\pi a^2 \int_0^1 v_0(\tfrac{1}{2} - z^2) C z \, dz. \tag{2.5.17}$$

Inserting the value of C from above, Taylor obtained

$$Q = -\frac{\pi a^4 v_0^2}{192 D} \frac{\partial C_{x_1}}{\partial x_1}. \tag{2.5.18}$$

If (2.5.18) is expressed in terms of the mean concentration C_m, one obtains, using the assumption that radial variations of C are small,

$$Q = -\frac{\pi a^4 v_0^2}{192 D} \frac{\partial C_m}{\partial x_1}. \tag{2.5.19}$$

It is therefore seen that C_m is dispersed relative to a plane which moves with velocity $\tfrac{1}{2}v_0$ as though it were being diffused by a process obeying the same law as molecular diffusion, but with a diffusivity coefficient D_1 where

$$D_1 = \frac{a^2 v_0^2}{192 D}. \tag{2.5.20}$$

The fact that no material is lost in the process is expressed by the continuity equation for C_m, viz.:

$$\frac{\partial Q}{\partial x_1} = -\pi a^2 \frac{\partial C_m}{\partial t}. \tag{2.5.21}$$

Substituting the value for Q, the equation governing longitudinal dispersion becomes

$$D_1 \frac{\partial^2 C_m}{\partial x_1^2} = \frac{\partial C_m}{\partial t}. \tag{2.5.22}$$

This equation can be applied to the case where dissolved material of uniform concentration C_0 is allowed to enter the tube at uniform rate at $x = 0$ starting at time $t = 0$. Initially the tube is assumed to be filled with pure solvent. One obtains (Von Rosenberg, 1956):

$$\frac{C}{C_0} = \frac{1}{2} + \frac{1}{2}\operatorname{erf}\left\{x_1\left[\frac{48D}{a^2v_0^2 t}\right]^{\frac{1}{2}}\right\} \quad \text{for} \quad x_1 < 0$$

$$\frac{C}{C_0} = \frac{1}{2} - \frac{1}{2}\operatorname{erf}\left\{x_1\left[\frac{48D}{a^2v_0^2 t}\right]^{\frac{1}{2}}\right\} \quad \text{for} \quad x_1 > 0$$

(2.5.23)

where

$$\operatorname{erf} z = 2\pi^{-\frac{1}{2}}\int_0^z e^{-z^2}\,dz. \tag{2.5.24}$$

Equation (2.5.23) signifies (Von Rosenberg, 1956) that the length of the concentration front—i.e., the length over which a certain fraction, say 80 per cent, of the total concentration change takes place—at any distance traversed is directly proportional to the square root of the velocity and inversely proportional to the square root of the diffusion coefficient. Also, at any given velocity, the length of the front increases as the square root of the distance traversed.

The above-outlined theory of Taylor (1953) is, as is evident from the exposition, an approximate one. Nevertheless, it seems to describe the experimental occurrences quite well (Taylor, 1953). There is no doubt, therefore, that it is at least qualitatively correct.

PART III

HYDROSTATICS IN POROUS MEDIA

3.1. Principles of Hydrostatics

The statics of fluids within porous media is governed by the same principles which obtain when fluids are confined in other types of vessels. Thus the basic theorems, such as that of Torricelli, will still be valid. There are, however, certain effects that are peculiar to fluids which are confined within porous media as a result of the proximity of the walls to practically all molecules of the fluid. These effects will be grouped for purposes of discussion into several classes, as follows:

(i) The statics of one fluid phase. The all-important phenomenon is that of adsorption of fluid particles at the surface of the pores. Such adsorption grossly alters the relationship between pressure and volume observed in bulk quantities of the fluids.

(ii) The statics of two phases of one fluid. The influence of the walls of the pores may cause some of the fluid to condense to form another phase. This second phase will cover the walls of the pores as a film of variable thickness. If it merges to form menisci, this phenomenon is known as "capillary condensation."

(iii) The statics of two different immiscible fluids. This subject is treated by the theory of capillaric forces acting at the interface between the two fluids. This is also the theory of the quasistatic displacement of one fluid by another from a porous medium. "Quasistatic" means that the displacement is taking place through a series of equilibrium-conditions without any proper dynamic effects occurring.

(iv) In a final section, the subject of wettability will be treated. The concept of wettability, actually, is nothing but a straightforward consequence of the displacement theory; nevertheless it merits special attention because of its widespread importance.

Hydrostatics in porous media has numerous applications to measurements of geometrical properties of porous media, which were mentioned in earlier sections of this monograph. These applications will be treated here in greater detail in connection with the basic underlying theories.

3.2. Adsorption of Fluids by Porous Media

3.2.1. Theory of Isotherms.
If an adsorbable fluid is confined within a porous solid, then the relationship between the pressure and the volume of the fluid is not the same as if the fluid were confined in a sphere of the same volume as the

pore space. A static interaction between the porous medium and the fluid is thus taking place. For a particular fluid and porous solid, one can find experimentally the relationship between fluid pressure and fluid density at a given temperature inside the porous medium. Such curves are referred to as adsorption (or desorption) isotherms.

Of the large number of papers on theories of adsorption, some of the most important will be reviewed here. A general presentation of adsorption phenomena has been given in a fundamental monograph by Brunauer (1943); less detailed reviews may be found in any text-book on colloid chemistry (see, for example, Hartman, 1947), as well as in articles by Cremer (1950) and Everett (1950).

Gibbs (Hartman, 1947) was probably the first to investigate the problem. Using general principles of thermodynamics he derived a relationship between the excess number ω of adsorbed molecules per unit area, the concentration C of molecules in the interior of the fluid, and the surface tension γ. This relationship is known as Gibbs' adsorption equation (for a modern write-up of the derivation of this equation see, for example, Glasstone, 1946, p. 1206). It is as follows:

$$\omega = (C/R\mathfrak{T})\partial\gamma/\partial C \qquad (3.2.1.1)$$

As usual, R signifies the gas constant and $\mathfrak{T}$ the absolute temperature.

The Gibbs equation reduces the phenomenon of adsorption to one of surface tension. The actual mechanics of the process is understood only if the dependence of γ on C is understood. This dependence has to be determined experimentally, which is equivalent to making an adsorption experiment and thus does not help the understanding of the mechanism. Nevertheless, the Gibbs equation restricts somewhat the general aspects of the adsorption process by admitting only possibilities that are thermodynamically sound. The Gibbs equation is usually applied to the analysis of the adsorption of a solute at the surface of a solution. It is, however, also applicable to the problem of adsorption on solids and in this respect it has been scrutinized by Bangham (1937).

Polanyi (1920) made a different attempt to understand the mechanism of adsorption. He developed a theory based on the assumption that the density of the adsorbed layer varies continuously with the distance from the surface of the adsorbent, owing to the action of an "adsorption potential." The adsorption potential is assumed to depend on the nature of the adsorbent as well as of the adsorbate, but not on the temperature and on the number of adsorbed molecules present. It should thus be possible to determine the adsorption potential from an isotherm; and, if the quantity of adsorbed matter is determined at a certain temperature as a function of the pressure, it should be possible to predict the quantity of adsorbed matter at any other temperature. Experimental checks of the Polanyi theory yielded discrepancies in many instances. Polyakov, Kuleshina and Neĭmark (1937), for example, stated that in the case of gels, the

Polanyi theory is applicable only for low adsorbent concentration, and yields incorrect results at higher ones.

Another theory of adsorption has been initiated by Langmuir (1916) who developed some consequences of the kinetic theory of gases. The latter theory provides a means of calculating the number of molecules of gas hitting a unit surface area in terms of the gas pressure, its molecular weight, and its temperature. Langmuir assumed that any molecules arriving at the surface of an adsorbent would stay there for a while such that a film of monomolecular thickness would develop. The molecules in the film, in turn, would re-emanate by a certain finite probability. The net effect of this process would be that a finite number of molecules would always be present at the surface of the adsorbent. This represents the phenomenon of adsorption.

Starting from these considerations, Langmuir (for a recent account of the mathematics see, for example, Hartman, 1947, p. 54) obtained an expression, in terms of the gas pressure p, of the quantity of gas ω which is adsorbed:

$$\omega = \frac{abp}{(1 + ap)}, \qquad (3.2.1.2)$$

or

$$\frac{p}{\omega} = \frac{p}{b} + \frac{1}{ab}. \qquad (3.2.1.3)$$

The last equation shows that one should obtain a straight line if p/ω is plotted against p. The quantities a and b are constants. They are constants because of the particular assumptions that Langmuir made, namely that adsorption could occur only in a single layer and that the energy of adsorption would be the same everywhere on the adsorbing surface. It follows therefore that the Langmuir equation should be valid only as long as there is free surface available to be taken up by molecules. After all free surface has been taken up, the adsorption process should come to an abrupt halt. This should manifest itself in a break of the adsorption isotherm.

Experiments showing that adsorption does not come to an abrupt halt in correspondence with the Langmuir assumptions, prompted Brunauer, Emmett and Teller (1938) to postulate the existence of adsorbed molecules in more than one layer. By a kinetic approach similar to that of Langmuir, they arrived at the following equation (the "BET-equation")

$$\frac{\beta}{V(1 - \beta)} = \frac{1}{V_m G} + \frac{\beta(G - 1)}{V_m G} \qquad (3.2.1.4)$$

where $\beta = p/p_0$ is the "relative" pressure (p is pressure, p_0 is vapour pressure of the bulk liquid) at which a particular volume V of gas (expressed in some standard measure) is adsorbed by the porous medium, V_m is a constant, namely the volume of gas required to cover the entire internal surface of the porous medium, and G is another constant. The two constants (V_m and G) can be determined for any particular system from a set of adsorption data by plotting $\beta/(V(1-\beta))$ against β.

The BET equation has been extended and scrutinized by many authors. Notably, the form (3.2.1.4) assumes the possibility of an infinite number of adsorbed layers. One can derive a similar equation that assumes only a finite number of such layers. The consequences of such a restriction to a finite number of layers have been discussed by Joyner, Weinberger and Montgomery (1945) and by Carman and Raal (1951). The basic assumptions of the multilayer adsorption theory have been analysed by McMillan and Teller (1951); the relationship of the BET equation to the Langmuir theory and earlier multilayer adsorption theories has been described by Keenan (1948).

A totally different approach to the theory of adsorption has been initiated by Harkins and Jura (1943, 1944c). These authors assumed that the same condition is applicable to the adsorption of gases on solids which had been found to govern the correlation between surface spreading force of adsorbed films with the area occupied per adsorbed molecule. Thus they were led to postulate that adsorption data should fit the following equation:

$$\log \beta = B - A/V^2 \qquad (3.2.1.5)$$

where A and B are constants and the other symbols have the same meaning as in equation (3.2.1.4). Furthermore, the constant A is expressible in terms of the internal surface area S, by the equation

$$S = cA^{\frac{1}{2}} \qquad (3.2.1.6)$$

where c is a constant.

It is quite obvious that the adsorption data of one experiment cannot fit both the BET *and* Harkins-Jura equations. Nevertheless, Emmett (1946) discussing this difficulty has shown that for quite a wide range of constants A, B, G, the two equations can be made to fit identical data very nearly. For the greater part of adsorption phenomena, the two equations seem therefore to be equally applicable.

3.2.2. Applications of Isotherms to Area Measurements. The theories of adsorption isotherms described above can be used to determine experimentally the internal area of porous media. A general review of the methods available has been given, for instance, by Emmett (1948).

It has been pointed out in section 3.2.1 that, according to the Langmuir theory, one should expect adsorption to cease after all surface available for occupation in a monomolecular layer by molecules of the adsorbate has been taken up. If the area occupied by one single molecule were known, one could then calculate the surface of the adsorbent knowing the amount of gas that can be totally adsorbed.

Unfortunately, adsorption does not come to a halt abruptly as expected by Langmuir since adsorbed molecules may exist in more than a single layer. It is, however, reasonable to expect that adsorption isotherms should contain a point that might correspond to the completion of the first layer of adsorbed molecules. This hypothetical point has been called "point B."

The method of selection of a "point B" has been described by Emmett and Brunauer (1937). The method makes use of the fact that most isotherms are characterized by a long linear part extending over a considerable portion of the pressure range, which may be thought to represent the building up of the second layer of adsorbed molecules. Therefore, the beginning of the long linear part would have to be identified with the "point B." However, the selection of a "point B" in this fashion is obviously somewhat arbitrary.

A better way to determine the volume of gas that makes up a single layer of adsorbate on the surface of the adsorbent is therefore obtained by the application of the BET equation. As outlined in section 3.2.1, the constant V_m in the BET equation represents precisely the sought-after quantity. If, in addition, the volume V_0 required to cover a unit area of adsorbent with a monomolecular layer of adsorbate, is known, the internal area S of adsorbent is given by

$$S = \frac{V_m}{V_0}. \qquad (3.2.2.1)$$

The method of BET is enjoying great popularity. It gives good results for "smooth" porous media, i.e., media without too narrow interstices. If the interstices are very narrow, the modification of the BET equation for a finite number of layers has to be used, and this greatly complicates matters.

Most of the area measurements reported earlier (in Table II) have been obtained by the BET method. Among the gases that have been used for surface measurements are nitrogen, oxygen, argon, hydrogen. The experiments have usually been performed at low temperature, around $-190°C$, i.e., at the temperature of liquefaction of the gases.

The method of BET still hinges on a determination of V_0, which cannot always be obtained very accurately. The theory of Harkins and Jura permits one to get around this difficulty. Their equations (3.2.1.5-6) can be applied immediately to surface-measurements if the constant c can be evaluated. In some ingenious experiments, Harkins and Jura (1944a, 1944b, 1945) were able to provide an independent means of evaluating c. They were able to measure the area of fine titanium oxide powder directly from measurements of the heat of wetting. Using this surface area, c was evaluated. For other solids, then, equation (2.3.1.6) could be applied directly by assuming the constant c as independent of the type of surface. In the cited papers, Harkins and Jura give several applications of their method to specific area determinations.

A comparison of the Harkins-Jura with the BET method has been made by Emmett (1946), who found satisfactory agreement.

3.3. Capillary Condensation of Fluids in Porous Media

3.3.1. Theory of Sorption Hysteresis. The theories of adsorption discussed so far presume that adsorption is an equilibrium process. However, it has been observed that under certain conditions there are phenomena of hysteresis. Such phenomena are usually explained in terms of capillarity effects or with more or less unspecific reference to phase transitions. The latter possibility is often referred to as "capillary condensation theory."

The various versions of the theory of capillary condensation are all based upon the assumption of Zsigmondy (1911) that vapours adsorbed on any porous solid are in equilibrium with a certain vapour pressure which is uniquely determined by the curvature of the menisci formed in the pores. One has thus two phases present in the fluid whose behaviour in relation to the porous medium is described by the relationships governing interfacial tension (see sec. 2.4.2). Since the interfacial tensions are subject to hysteresis, the behaviour of the capillary-condensed liquid will exhibit the same phenomenon. The original theory of Zsigmondy has been modified and improved by Foster (1932) and by Broad and Foster (1945). These authors suggested that hysteresis could also be due to a delay in the formation of the capillaric menisci. This implies that the equilibrium postulated by Zsigmondy might not be reached instantaneously, but only after a certain time-lag.

In any case, the capillary condensation theory is founded upon the assumption that, in a capillary, the saturation vapour pressure of a liquid is reduced from, say, p_0 to, say, p. The ratio p_0/p is termed "relative pressure." Assuming zero contact angle and a circular capillary of radius a, the relative pressure is given by the following equation established by Kelvin:

$$\log \text{nat} \, (p_0/p) = 2\gamma M/(\rho a R \mathfrak{T}). \qquad (3.3.1.1)$$

Here, γ is, as usual, the surface tension, M the molecular weight, ρ the density, $\mathfrak{T}$ the absolute temperature, and R the gas constant. The Kelvin equation describes matters correctly if it is assumed that (i) all "adsorption" is entirely due to capillary condensation, (ii) "adsorbate" densities equal bulk densities, (iii) differences of pore shape from circular can be ignored, and (iv) the validity of the Kelvin equation, including constancy of γ and p, is unimpaired at low values of a (Carman, 1951).

Hirst (1947) has shown that the capillary condensation theory may account for sorption hysteresis without the assumption of hysteresis in interfacial tension—for a reason other than an assumed time-lag in the formation of the menisci. If condensation takes place in a cylindrical cavity with thin isotropic walls, the radius of the capillary will be altered owing to the pressure of the liquid film. As a result of the change of radius, the equilibrium pressure over the liquid film is altered and hysteresis is observed.

In this connection, investigations by Shull (1948), Barrett, Joyner and Halenda (1951), and Carman (1951) are concerned with the fact that the first condition basic for the validity of the Kelvin equation, *viz.*, assuming that "adsorption" is due *only* to capillary condensation, is really quite unreasonable, since formation of truly adsorbed layers certainly begins before capillary condensation is manifested and must continue at the surface of the unfilled capillaries. To obtain a realistic theory, a correction must therefore be introduced for the adsorbed layers. The details of this are, however, beyond the scope of this monograph.

The sorption hysteresis has also been attributed to an "ink bottle effect" instead of to hysteresis in γ (Katz, 1949). This is a blockage effect of the smaller capillaries which may stay filled during desorption and prevent the connecting larger ones from being properly emptied. It appears thus that in any interpretation, the concept of "capillary condensation" leads to sorption hysteresis.

Most of the above-mentioned investigations discuss some experimental tests. In addition, Carman and Raal (1951) provided direct evidence of capillary condensation and of blockage of capillaries with adsorbed layers by comparing the adsorption on a given surface both as a free surface and as the internal surface of a porous medium. Polyakov, Kuleshina and Neĭmark (1937) made some experimental tests and showed that at high adsorbent concentration (i.e., narrow pore spaces) the capillary condensation theory may give satisfactory results. However, Bond, Griffith and Maggs (1948) conducted a set of experiments to determine the actual state of aggregation of the "capillary-condensed" film of "liquid." They pointed out that certain aspects of the behaviour of the adsorbed "film" stand in very sharp contrast to the behaviour of bulk quantities of such liquid. In view of these experiments it must be conceded that sorption hysteresis is not yet completely understood.

3.3.2. Determinations of Pore Size Distribution.

The various theories of "adsorption" (including capillary condensation) can be used to determine the pore size distribution. Naturally, as was outlined in section 1.1, the meaning of the term "pore size distribution" is in most publications not very well defined and it is therefore usually a little vague. Reported determinations of pore size distribution are therefore not always entirely satisfactory from a quantitative standpoint. However, a good qualitative indication of the nature of the pore space involved is, nevertheless, usually obtained.

There are various methods for determination of pore size distribution based upon adsorption isotherms. Foster (1938) discerns three classes:

(a) Methods based upon the capillary condensation theory. It follows from the Kelvin equation that the radius a of the smallest pores that may be invaded by a given liquid at constant temperature, is solely a function of the pressure p, provided that the surface tension γ and the density ρ are assumed as unaffected by a. It therefore also follows that, after an "adsorption" isotherm for a porous medium has been determined, it is merely necessary to convert values of p to corresponding values of a to obtain the pore size distribution of the porous medium. Actually, corrections should be made for the simultaneous occurrence of multilayer adsorption with the capillary condensation, as was stated above. The Kelvin equation by itself, without such a correction, is somewhat restricted in its accuracy for the purpose of pore size distribution determination, although it has been applied to such measurements with good qualitative success.

(b) Methods based upon the determination of the internal surface. The radius a of a straight circular cylinder can be calculated from the ratio of its volume V to its wall-surface S as follows

$$a = \frac{2V}{S}. \tag{3.3.2.1}$$

If this ratio is assumed to apply approximately to the pores of a porous medium (Foster, 1934; Emmett and de Witt, 1943), then internal surface values of porous media can be converted to yield "average" pore radii.

(c) Methods yielding approximate estimates based on a consideration of the shape of adsorption isotherms. If adsorption practically ceases after what may be assumed to be the formation of a monolayer, one may infer that the pore radius is of the order of 1–2 molecular diameters. It can therefore be estimated (Foster, 1945).

The various methods of pore size distribution determination outlined above have been applied to a great variety of substances. These methods have been compared (Joyner et al., 1951) with the more common mercury-injection method (see sec. 3.4.3). It was found that there is, in view of the general vagueness of the notion of pore size distribution, a satisfactory qualitative agreement.

3.4. The Quasistatic Displacement of One Liquid by Another in a Porous Medium

3.4.1. Theory of Capillary Pressure.
Let us now consider the hydrostatics of two immiscible fluids or phases that may exist simultaneously in a porous medium. In general, one phase will wet the solid. Experimental investigations have shown (Versluys, 1917, 1931) that there are three general types of occurrence of one of the two phases, or régimes of saturation with that phase:

(i) Saturation régime. The porous medium is completely saturated with one phase.

(ii) Pendular régime. The porous medium has the lowest possible saturation with one phase. This phase occurs in the form of pendular bodies throughout the porous medium. These pendular bodies do not touch each other so that there is no possibility of flow for that phase. A drawing of the pendular régime in the case of an idealized porous medium (consisting of spheres) is shown in figure 6a.

(iii) Funicular régime. The porous medium exhibits an intermediate saturation with both phases. If the pendular bodies of the pendular régime expand through addition of the corresponding fluid, they eventually become so large that they touch each other and merge. The result is a continuous network of both phases across the porous medium. It is thus possible that simultaneous flow of both phases occurs along what must be very tortuous (funicular) paths.

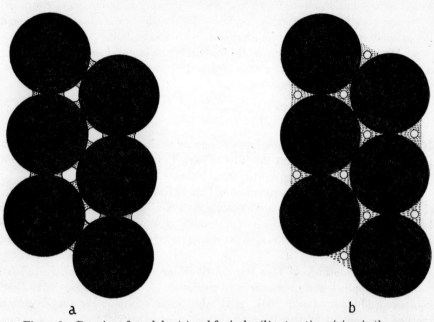

a   b

Figure 6. Drawing of pendular (a) and funicular (b) saturation régime in the case of an idealized porous medium consisting of packed spheres. (After Versluys, 1931.)

A drawing of the funicular saturation régime for an idealized porous medium (consisting of spheres) is shown in figure 6b.

The entrance of fluid into a small pore against another liquid is opposed (or helped) by the surface tension (Washburn, 1921). The combined actions of all these forces are such that, at any given relative saturation, a certain pressure differential in the displacing phase *versus* the displaced phase will have to be maintained to create equilibrium. This pressure is called "capillary pressure" (p_c); it is a function of the saturation s:

$$p_c = p_c(s) = p\text{ (Phase 1)} - p\text{ (Phase 2)} \qquad (3.4.1.1)$$

Capillary pressure, like surface tension, exhibits the phenomenon of hysteresis.

One can obtain an idea about the origin of capillary pressure by noting that the interfacial area between the fluids will change. The interfacial tension

γ_{12} between the two fluids is, by definition, the free surface energy F per unit interfacial area:

$$\gamma_{12} = dF/d\Sigma,$$

where Σ is the two-fluid interfacial area per unit pore volume. Following Leverett (1941), we *assume* for the time being the existence of capillary pressure. Then, the difference in free surface energy between two saturation states (denoted by s_A and s_B) in the same porous medium is

$$\Delta F = -\int_{s_A}^{s_B} p_c ds;$$

and therefore one has:

$$\Sigma_B - \Sigma_A = (1/\gamma_{12})\int_{s_A}^{s_B} p_c ds. \quad (3.4.1.2)$$

Thus, if there occurs a change of interfacial area the capillary pressure on the right hand side of equation (3.4.1.2) *must actually exist* and be different from zero.

A further qualitative explanation, as discussed for example, by Adam (1948), of the existence of capillary pressure is obtained by visualizing the porous medium as consisting of an assembly of capillaries of diameters δ_c and a frequency $\alpha(\delta)$ (see sec. 1). In a single capillary, the curvature r of the interfacial surface gives rise to a pressure differential across the interface equal to

$$p_c = 2\gamma/r.$$

The radius of curvature of the meniscus is equal to $\delta_c/(2 \cos \theta)$ so that one has for a single circular capillary (θ is as usual the contact angle):

$$p_c = 4\gamma \cos \theta/\delta_c. \quad (3.4.1.3)$$

The combined pressure in all those capillaries will give rise to the capillary pressure. The hysteresis between advancing and receding contact angles explains, then, why capillary pressure will also exhibit the effect of hysteresis.

Capillary pressure is specific to the nature of the two fluids involved. However, if no further specification is made, it is usually understood that the displaced "fluid" is the vacuum.

If the displaced "fluid" is the vacuum, and an external pressure p_c is applied in a non-wetting fluid, then all capillaries with a diameter larger than δ_c will be totally filled. The connection between saturation and this capillary pressure is thus given by the equation:

$$s = \int_{\delta\,=\,4\gamma \cos \theta/p_c}^{\infty} \alpha(\delta) d\delta. \quad (3.4.1.4)$$

If the capillaries are not circular, the equation for the capillary pressure has to be generalized by replacing $2/r$ by $1/r_1 + 1/r_2$

$$p_c = \gamma \left(\frac{1}{r_1} + \frac{1}{r_2} \right) \qquad (3.4.1.5)$$

where r_1 and r_2 are the principal radii of curvature of the meniscus. Schultze (1925a, 1925b) has shown that the capillary pressures for such capillaries under the assumption of zero contact angle are approximately given by the equation

$$m = \frac{\gamma}{p_c} \quad \text{or} \quad \frac{1}{m} = \frac{1}{r_1} + \frac{1}{r_2}, \qquad (3.4.1.6)$$

where m is the ratio of volume to surface of the capillary. The latter is sometimes also called "hydraulic radius" of the capillary. A list of comparative values to test equation (3.4.1.6) is given in table III (after Carman, 1941).

TABLE III

List of Comparative Values to show Equivalence of the Reciprocal Hydraulic Radius ($1/m$) with the Reciprocal Mean Radius of Curvature ($1/r_1 + 1/r_2$) in a Capillary (r_i is the Radius of the Inscribed Circle). (After Carman, 1941)

Cross-section		$1/r_1 + 1/r_2$	$1/m$
Circle		$2/r$	$2/r$
Parallel plates		$1/b$	$1/b$
Ellipse	$a:b = 2:1$	$1.50/b$	$1.54/b$
	$a:b = 5:1$	$1.20/b$	$1.34/b$
	$a:b = 10/1$	$1.10/b$	$1.30/b$
Rectangle		$1/a + 1/b$	$1/a + 1/b$
Equilateral triangle		$2/r_i$	$2/r_i$
Square		$2/r_i$	$2/r_i$

If the pore-openings are not of straight capillaric form, formula (3.4.1.5) is still a valid expression for the capillary pressure. In order to obtain a theoretical relationship between saturation and capillary pressure for a porous medium, the crucial point is obviously to find an analytical expression for the average interfacial curvature as a function of saturation. This is a very difficult proposition.

Often, it is sufficient to consider the "capillary rise" of a wetting fluid against another in a porous medium. Such a rise will reach a height h at which the saturation of the wetting fluid will drop sharply. The height h is, of course, not really exactly defined, but it serves as a satisfactory working assumption in many applications. A theoretical estimate of this height is given below (following Carman, 1941).

As equation (3.4.1.6) can give a reasonably accurate correlation of capillary pressure in non-circular capillaries, it will likely also be applicable to the capillary channels in a porous medium. This idea amounts to nothing less than a generalization of the relationship (3.4.1.6) which was obtained from a particular model of straight capillaries, to arbitrary porous media. Thus denoting by P the fractional porosity of the porous medium, and by S its specific surface area, one has

$$\Delta \rho g h = \gamma S/P, \qquad (3.4.1.7)$$

where $\Delta \rho$ is the density difference of the displacing *versus* the displaced fluid. If the porous medium is unconsolidated, and if S_0 is the particle surface per unit *solid* volume, then $S = (1 - P)S_0$. For spherical particles of uniform diameter δ, S_0 is equal to $6/\delta$, and therefore one has

$$\Delta \rho g h = \frac{6(1-P)\gamma}{P\delta}. \qquad (3.4.1.8)$$

This last equation may be used as a general relationship for arbitrary porous media although it was derived using a specific model. Data substantiating it can be taken from the experimental work of Atterberg (1918), Smith, Foote and Busang (1931), and Hackett and Strettan (1928) (cf. Carman, 1941).

The theory of Carman is concerned with the notion of "capillary rise" only. A consistent theory of capillary pressure in porous solids should provide an explanation of the fundamental relationship between saturation and capillary pressure (or interfacial curvature). To date, this does not seem to have been obtained. Therefore, Leverett (1941) chose a semi-empirical approach by showing that a dimensionless expression, *viz.*:

$$J = \frac{p_c}{\gamma} \left(\frac{k}{P}\right)^{\frac{1}{2}} = J(s) \qquad (3.4.1.9)$$

k = permeability

which is termed the "Leverett function," can be plotted against the saturation s of the wetting fluid and that the data for a number of unconsolidated porous media fall satisfactorily near two curves, one for imbibition of the wetting fluid and the other for drainage. In (3.4.1.9), k is the permeability (see sec. 4.2) of the medium. Typical examples of some Leverett functions are given in figure 7.

The form of the Leverett correlation may be derived from either of two assumptions (Leverett, 1941): (i) that the pressure at which a definite wetting fluid saturation is found at equilibrium is inversely proportional to an "equivalent circular diameter" of the pores calculated from the porosity and permeability of the medium (cf. sec. 4), inversely proportional to the density difference, and directly proportional to the interfacial tension; or (ii) that the interfacial surface area between the two fluids is, at a given water saturation, a definite fraction of the total surface of the porous medium itself. Leverett does not shed any light, however, upon the particular form of the curves $J = J(s)$ which, from a physical standpoint, is unaccounted for. Nevertheless, the assumption of

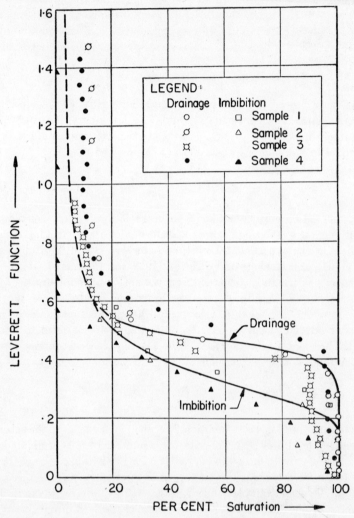

Figure 7. Typical examples of some Leverett functions. (After Leverett, 1941.)

certain shapes of J-curves for certain groups of porous materials, serves well as a working assumption for applications.

3.4.2. Experiments on Capillary Pressure. A capillary pressure curve of a porous medium is usually obtained by enclosing the latter in a cell and displacing one fluid with another slowly by applying pressure in the displacing phase.

Unless the displaced phase is the vacuum or else a highly compressible fluid, an outlet has to be provided for it; this can be done by applying a semipermeable medium which is impermeable to the displacing phase.

The saturation at any pressure can be determined by measuring the amount of displacing fluid that entered the porous medium, by electrical measurement (Martin et al., 1938), or by X-ray techniques (Boyer, Morgan and Muskat, 1947; Morgan, McDowell and Doty, 1950; Laird and Putnam, 1951). Capillary pressure curves have been obtained in large numbers by workers in the oil industry, with oil as displaced and water or gas as displacing fluid. Typical experiments have been reported by Hassler and Brunner (1945), Calhoun, Lewis and Newman (1949), Powers and Botset (1949), Rose and Bruce (1949), Stahl and Nielsen (1950), Slobod, Chambers and Prehn (1951), and Dunning et al. (1954). The general shape of the curves bears out that often no displacement occurs until a certain minimum "displacement pressure" (Whitney and Bartell, 1932) is reached in the displacing phase, and that always a certain residual saturation of the displaced fluid remains no matter how great a pressure is applied in the displacing fluid. A typical example of a capillary pressure curve, obtained by displacing water by oil from a porous medium, is shown in figure 8.

Capillary pressure curves are very sensitive to the state of the internal surface of the porous medium owing to the dependence of p_c on γ and the contact angle. Apart from the fact of their showing much hysteresis, a phenomenon which is aggravated by the possibility of physical blockage during a depletion experiment, it is often difficult even to repeat experiments to obtain the same curves: the preferential wettability (i.e., the contact angle) may have been changed by the previous experiment. Thus, it has been proposed to duplicate the process performed by nature, which consisted of displacing water by oil in certain rock strata down to a residual water saturation (the latter is always found in oil-bearing formations), in the laboratory (Bruce and Welge, 1947; Thornton and Marshall, 1947), which would give a means of determining the "connate" water saturation of an oil field *a posteriori*. It soon became evident, however, that this method does not work (Yuster and Stahl, 1948) simply because in the laboratory experiment the contact angle is different from that encountered in the original natural process. Moreover, it is also impossible to correlate capillary pressure curves that have been obtained with different fluids (Brown, 1951), although one might think from the formal appearance of p_c (equation 3.4.1.5) that this should be possible (Purcell, 1949).

3.4.3. Capillary Pressure and Pore Size Distribution. Equation (3.4.1.4) states a connection between the void size distribution $\alpha(\delta)$ of a certain type of capillaric model and the capillary pressure curve where the displacing fluid does not wet the medium and is entering it against vacuum. One might propose, therefore, to use this equation to obtain $\alpha(\delta)$ from a capillary pressure curve for a given porous medium.

It should be noted, however, that the capillaric model upon which equation (3.4.1.4) is based, is a far cry from an actual porous medium. It is therefore not to be expected that, if $\alpha(\delta)$ be calculated from such capillary pressure curves,

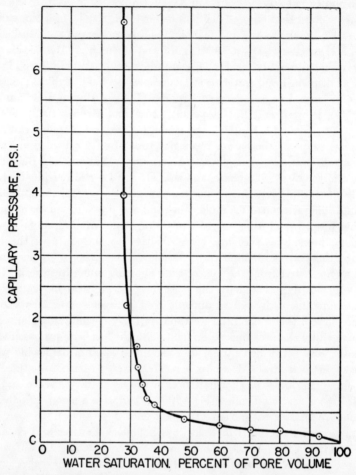

Figure 8. Typical capillary pressure curve, obtained by displacing water by oil from a porous medium. Note the residual water saturation in spite of the application of very high capillary pressures. (After Bruce and Welge, 1947.)

one will actually obtain the pore size distribution as defined in section 1. In this instance Meyer (1953) at least attempts a correction for those large pores that are connected to the "outside" only by small ones and thus will not be filled at the corresponding pressure. However, most of these "pore size distribution determinations" have been made without such corrections, which, at best, gives a good qualitative indication of the nature of the pore spaces involved.

§3.4   QUASISTATIC DISPLACEMENT IN A POROUS MEDIUM   51

The capillary pressure method of pore size distribution determination usually takes the form of evacuating the porous medium in a cell, and then of injecting non-wetting fluid—usually mercury—at various pressures. The method goes back to Washburn (1921) and has later been applied on various occasions by Henderson, Ridgway and Ross (1940), and Loisy (1941). It was finally greatly

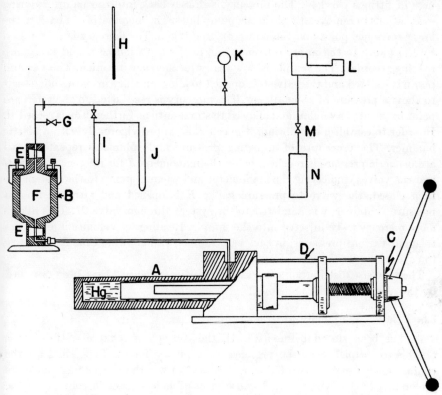

Figure 9. A typical apparatus for determination of pore size distribution by the mercury-injection method. Courtesy of Imperial Oil Limited.

popularized by Ritter and Drake (1945), (also Drake, 1949) after whom a great number of pore size distribution determinations were performed by the mercury injection method.

A typical apparatus for routine pore size distribution determination by the mercury injection method is shown in figure 9. It consists, essentially, of a cell F to take up the sample, and a pumping mechanism A to inject the mercury. In detail, the experiment proceeds as follows. Prior to the use of the apparatus, the internal volume of the cell F is measured by the use of the calibrated pump A and measured steel blanks. The sample is then placed in the cell F and the lid tightened down. With the cell at atmospheric pressure, the mercury level is brought up to the lower reference point in the lucite window E and the pump reading recorded. The vacuum pump L is turned on and a vacuum of less than 0.1 mm. as read on the closed arm manometer I is obtained. Pumping is continued for one hour to remove absorbed vapours, and the tightness of fittings checked. The mercury is drawn back into the pump A during evacuations to remove any air in the pump lines. The mercury is run back to the lower reference point and the pump reading taken. The mercury is then raised by the pump to the upper reference line in the top window E and the pump reading recorded. The sample is now ready for mercury injection. The vacuum pump is isolated and the valve G is opened to allow mercury in the manometer J to show a pressure of about 10 cm. The mercury is readjusted to the reference point in the upper window and allowed to stand until no further drop is recorded. In order to establish equilibrium, the cell B is tapped vigorously with a plastic hammer. The procedure of adjusting pressure and volume is repeated until atmospheric pressure is reached, using the manometer J for reference. Valve G and the valves connecting the manometer and vacuum pump to the system are then closed, the valve to pressure gauge K is opened and nitrogen from a pressure container N is admitted to the system through valve M. This allows more mercury to be injected into the sample. Readings of volumes injected are taken at various pressures, in each case tapping and allowing time for mercury to enter the pores.

The pore size distribution curves shown earlier in figure 2 have been obtained by the above method.

3.5. Relative Wettability

As has been stated before (sec. 1), the concept of relative wettability is another consequence of the presence of capillary forces. Depending on the contact angle between two fluids, one is said to wet the porous medium more preferentially than the other. Because of possible hysteresis in contact angles, the relative wettability of a porous medium by two fluids may also be hysterical. The determination of the relative wettability of a porous medium by two fluids can be performed simply by measuring the mutual displacement pressures (Bartell and Osterhof, 1927).

Owing to energy changes during the process of wetting of a porous solid, a certain amount of heat is liberated. This "heat of wetting" can be used for internal area determinations as it should be proportional to the area which is being wetted by the fluid.

The heat of wetting due to the complete soaking of a porous medium by a wetting fluid can be estimated as follows (Wenzel, 1938).

When the liquid spreads over an area S_1, a solid-air interface of that area is destroyed and thus an energy of $S_1 \gamma_{SA}$ is gained, where γ_{SA} is the solid-air interfacial tension (see sec. 2.4.2). At the same time, a solid-liquid interface of the same area S_1 is formed, which means that an energy $S_1 \gamma_{SL}$ has to be expended. In addition, a liquid-air interface of area $S_2 (\neq S_1)$ is formed, requiring the additional expenditure of an amount of energy equal to $S_2 \gamma_{LA}$. Thus, the total amount of energy gained in the process (i.e., the heat of wetting) is

$$W = S_1 \gamma_{SA} - S_1 \gamma_{SL} - S_2 \gamma_{LA}. \tag{3.5.1}$$

Introducing the value (2.4.2.2) for the contact angle θ, this becomes:

$$W = \gamma_{LA}(S_1 \cos \theta - S_2). \tag{3.5.2}$$

In soaking a porous medium, the solid-liquid interface S_1 will be much larger than the liquid-air interface S_2, and therefore S_2 can be neglected. Furthermore, S_1 becomes equal to the total internal surface area S of the porous medium and the expression for the heat of wetting thus becomes:

$$W = \gamma_{LA} S \cos \theta. \tag{3.5.3}$$

This is the required connection between the heat of wetting W and the internal surface S of a porous medium.

PART IV

DARCY'S LAW

4.1. The Flow of Homogeneous Fluids in Porous Media

The present part of this monograph has as its subject one of the aspects of the flow of homogeneous fluids in porous media, *viz.*, that in which the law of Darcy is valid. In the course of this part, we shall first discuss Darcy's experiment as it was originally performed and then enumerate further experiments made to elucidate the laws of flow through porous media.

The experiments of Darcy and of later workers suggest a law of flow, which, however, is not uniquely defined. Therefore, there is considerable ambiguity in postulating a differential equation which would be equivalent to the results of those experiments. In fact, the differential equation which is now commonly called "Darcy's" law, is not an equivalent expression for the findings of Darcy, although these do follow from it. However, they would equally well follow from other types of differential equations. This is especially true if generalizations of Darcy's law are attempted with respect to anisotropic and compressible porous media. Some discussion will be devoted to this subject later in this review.

Another section of the present part will be devoted to methods of measuring the coefficient occurring in Darcy's law, *viz.*, to methods of measuring the "permeability." The literature on permeability determination is extremely prolific as this quantity is important in many applications. The differential form of Darcy's law is also basic to the theory of the process of filtration. This fourth part will therefore be concluded by a review of some aspects of filtration.

It is to be expected that there are limitations of Darcy's law. Indeed, such limitations occur generally at high and at low flow rates, as well as in relation to various other effects. However, the range of validity of Darcy's law and its limitations will be discussed in a later part of this book devoted to general flow equations.

The subject of homogeneous flow through porous media has many technical and engineering applications. Therefore, general accounts of the theory may be found in a number of text-books on such subjects. The only books that deal specifically with flow through porous media are those by Muskat (1937), Leïbenzon (1947), and Polubarinova-Kochina (1952) which have been mentioned in the Introduction. Of the authors who deal with applications which are based on the theory of flow, and which therefore contain rather long reviews of the latter, the following may be mentioned. Forchheimer (1930) and Nemenyi (1933) wrote books on general hydraulics, and Muskat (1949) and Calhoun (1953) on the principles of oil production. Dickey and Bryden (1946) discussed the flow through porous media in connection with filtration, and Leva et al. (1951) wrote of the flow through packed towers. Most numerous are text-books on soil mechanics and ground water flow, which contain chapters on flow through porous media. Such books have been written by Blank (1939), Keen (1931), Dachler (1936), Tsunker (1937), Baver (1940), Agadzhanov (1947), and Terzaghi (1951). A book by Dalla Valle (1943) on micrometrics also contains a chapter on flow through beds of particles.

Of general reviews of the subject that appeared in journals, the following may be mentioned. Seepage and flow through beds have been reviewed by Hubbert (1940), Meinzer and Wenzel (1949), Hancock (1942), and Polubarinova-Kochina and Falkovich (1947). The last-mentioned article is particularly useful as it gives a seven-page bibliography on recent Russian work. Further reviews have been made by Slater (1948), Verschoor (1950a, 1950b), and Wicke and Brötz (1952).

The theory of flow through porous media has been reviewed with particular reference to the process of filtration by Nelson-Skornyakov (1949), Silverman (1950), Rietema (1951), Waeser (1951), and Wilson (1951). The annual reviews in *Industrial and Engineering Chemistry* on advances in filtration theory also merit mention. Similar reviews on fluid dynamics in the same journal also contain much information on the advances in the theory of flow through porous media.

4.2. Experimental Investigations

4.2.1. The Experiment of Darcy.
The theory of laminar flow through homogeneous porous media is based on a classical experiment originally performed by Darcy (1856). A modern discussion of this experiment may be found in an article by Hubbert (1940).

A schematic drawing of Darcy's experiment is shown in figure 10. A homogeneous filter bed of height h is bounded by horizontal plane areas of equal size A. Both these areas are congruent so that corresponding points could be connected by vertical straight lines. The filter bed is percolated by an incompressible liquid. If open manometer tubes are attached at the upper and lower boundaries of the filter bed, the liquid rises to the heights h_2 and h_1 respectively above an arbitrary datum level. By varying the various quantities involved, one can deduce the following relationship:

$$Q = \frac{-KA(h_2 - h_1)}{h}. \qquad (4.2.1.1)$$

Here, Q is the total volume of fluid percolating in unit time, and K is a constant depending on the properties of the fluid and of the porous medium. The relationship (4.2.1.1) is known as *Darcy's law*. The minus-sign in the expression for Q indicates that the flow is in the opposite direction of increasing h.

Darcy's law can be restated in terms of the pressure p and the density ρ of the liquid. At the upper boundary of the bed (elevation above the datum level

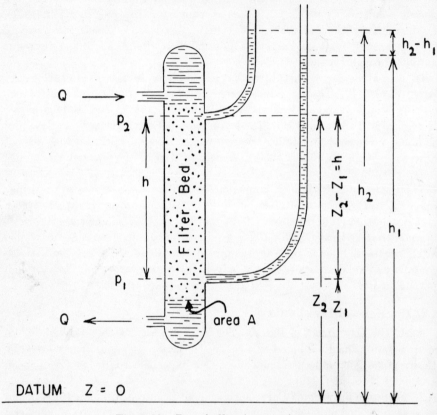

Figure 10. Darcy's filtration experiment.

denoted by z_2), the pressure is $p_2 = \rho g(h_2 - z_2)$, and at the lower boundary (elevation above datum level denoted by z_1), the pressure is $p_1 = \rho g(h_1 - z_1)$. Inserting this statement into (4.2.1.1), one obtains (as $z_2 - z_1 = h$)

$$Q = - KA[(p_2 - p_1)/(\rho g h) + 1];$$

or, upon introduction of a new constant K',

$$Q = - K'A(p_2 - p_1 + \rho g h)/h. \qquad (4.2.1.2)$$

The equations (4.2.1.1) and (4.2.1.2) are equivalent statements of Darcy's law.

The validity of Darcy's law has been tested on many occasions (e.g., Vibert 1939; Iwanami 1940; LeRosen 1942). Thus, it has been shown that it is valid for a wide domain of flows. For liquids, it is valid for arbitrary small pressure differentials (Meinzer and Fishel, 1934; Fishel, 1935; Meinzer, 1936; and Schweigl and Fritsch, 1941). It has also been used to measure flow rates by determining the pressure drop across a fixed porous plug (see Souers and Binder, 1952). For liquids at high velocities and for gases at very low and at very high velocities, Darcy's law becomes invalid, as will be discussed later.

4.2.2. The Permeability Concept. The law of Darcy in its original form is rather restricted in its usefulness. The first task is to elucidate the physical significance of the constant K'. The constant is obviously indicative of the permeability of a certain medium to a particular fluid. It depends on the properties of both the medium and the fluid. Before about 1930, it was often named "permeability-constant"; its dimensions are $M^{-1} L^3 T^1$, with mass (M), length (L), and time (T) as fundamental dimensions.

A constant of the type K', however, is not very satisfactory because one would like to separate the influence of the porous medium from that of the liquid. Nutting (1930) had already stated that one should have

$$K' = k/\mu, \qquad (4.2.2.1)$$

where μ is the viscosity of the fluid and k the "*specific* permeability" of the porous medium. But this relationship was not generally accepted until it was popularized by Wyckoff et al. (1933). The verification of it consists in the innumerable successful determinations of permeability that have been performed upon its basis.

The dimension of (specific) permeability is a length squared, which suggests that the natural permeability unit in the c.g.s. system should be cm^2. Unfortunately, this unit has been adopted only by some physicists and chemists. In most branches of applied science, some other unit has been adopted specific to that branch. Thus, the oil industry, for example, uses the "Darcy," with

$$1 \text{ Darcy} = 9.87 \times 10^{-9} \text{ cm}^2. \qquad (4.2.2.2)$$

In groundwater hydrology, it is customary to represent permeability in terms of the velocity of the percolating water per unit drop of hydraulic head. According to country and inclination, this velocity may be expressed in terms of any combination of feet, centimetres, hours, minutes, or seconds. One has, for example:

$$1 \text{ cm/sec for water} = 1.02 \times 10^{-5} \text{ cm}^2. \qquad (4.2.2.3)$$

In order to compare permeabilities for various substances in various branches of applied science, it is best to transform everything into c.g.s. units.

Once the permeability concept has been defined, further experiments have to be directed towards determining the range of actual constancy of this permeability for a given porous medium. In other words, one has to determine the dependence of the permeability on external conditions of the porous medium.

It would be expected that the permeability is dependent upon external stresses that might be impressed upon the porous medium assuming that the latter is compressible. Indeed, this is a well-known fact in the theory and

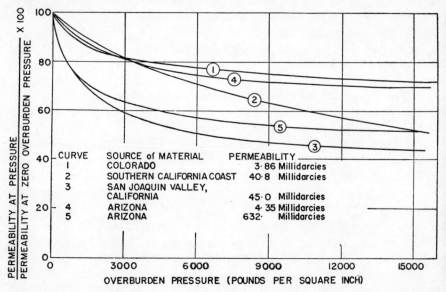

Figure 11. Curves showing the dependence of (specific) permeability on total external stress (i.e., overburden pressure). (After Fatt and Davis, 1952.)

practice of filtration. Thus, Secchi (1936) has made experiments to determine the dependence of permeability of a filter on the external pressure and has been able to show, not only that there is such a dependence, but also that it is subject to hysteresis. Another series of experiments has been published by Ruth (1946). Tiller (1953) showed that the dependence of the permeability on the fluid pressure p and the total pressure $\mathscr{T}$ (see sec. 1.6) can be represented as follows:

$$k = K(\mathscr{T} - p)^{-m}. \qquad (4.2.2.5)$$

This relationship was deduced from largely empirical investigations combined with notions of the Kozeny theory (see Part VI of this review). K and m are constants that have to be determined experimentally. The relationship is valid only if $(\mathscr{T} - p)$ is larger than some (experimentally determined) limiting value. Fatt and Davis (1952) have made similar investigations of the dependence of permeability on stresses in connection with the study of petroleum well cores. Another set of experiments has been performed by Grace (1953). In figure 11 we show some curves demonstrating the dependence of permeability on the total external stress.

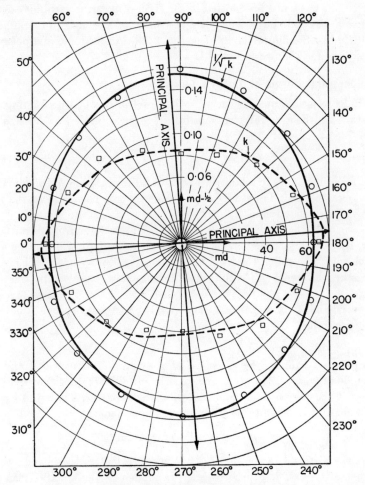

Figure 12. Polar permeability diagram. The actually measured values (squares) fit on a complicated curve (dashed) which corresponds to an ellipse (solid) if $k^{-\frac{1}{2}}$ were drawn instead of k. The circles correspond to the values of $k^{-\frac{1}{2}}$ as measured; it is seen that the fit of these values with an ellipse is excellent. This substantiates the tensor theory of permeability. (After Scheidegger, 1954.)

There is also the possibility of a directional variation of permeability. Thus, if a cube is cut out of a macroscopically homogeneous piece of rock, the permeability may not be the same across all the faces. This effect has actually been observed by Sullivan (1941), Pressler (1947), Johnson and Breston (1948), and Griffiths (1950). In this connection, a very valuable set of experiments has also been made by Johnson and Hughes (1948): the directional permeability was measured in intervals around 180 degrees in sections of a number of natural rock pieces. The results are plotted in the form of polar permeability diagrams. Most polar diagrams have somewhat the shape of an ellipse; others, however, are a little like the figure "eight" (see sec. 4.3.3). An example of a polar permeability diagram is shown in figure 12.

4.3. Differential Forms of Darcy's Law

4.3.1. Isotropic Porous Media.
The law of Darcy when accounting for the separation of the general constant into "permeability" and "viscosity," is expressible as follows:

$$q \equiv Q/A = -(k/\mu)(p_2 - p_1 + \rho g h)/h. \qquad (4.3.1.1)$$

In this form, it applies to a horizontal bed of finite thickness h, being percolated by an incompressible liquid of (constant) density ρ. This form of the law has only a very restricted use.

The aim will, therefore, be to express (4.3.1.1) in differential form. Unfortunately, there is no unique way of doing this. Naturally, q will become a vector $\mathbf{q}$ which might be called the local "filter-velocity," or "seepage-velocity," and the pressure difference in (4.3.1.1) must be somehow expressed by the pressure gradient.

A first possibility is suggested from (4.3.1.1) by letting h become infinitesimal. One then obtains:

$$\mathbf{q} = -(k/\mu)(\text{grad } p - \rho \mathbf{g}) \qquad (4.3.1.2)$$

where $\mathbf{g}$ is a vector in the direction of gravity (i.e., *down*ward) and of the magnitude of gravity. However, the experiment of Darcy does not tell us what happens if the permeability and viscosity are variables. Thus, the coefficient might equally well have to be taken into the gradient:

$$\mathbf{q} = -\text{grad } (kp/\mu) + k\rho \mathbf{g}/\mu \qquad (4.3.1.3)$$

The first possibility (equation 4.3.1.2) is equivalent to the introduction of a force potential ϕ (if it can be defined) inasmuch as the equation can be rewritten in an equivalent manner as follows (see Hubbert, 1940):

$$\phi = gz + \int_{p_0}^{p} dp/\rho(p), \qquad (4.3.1.4a)$$

$$\mathbf{q} = -(k\rho/\mu) \text{ grad } \phi, \qquad (4.3.1.4b)$$

where z denotes the vertical co-ordinate. Correspondingly, in the second possibility, equation (4.3.1.3) is equivalent to the introduction of a velocity-potential (if it can be defined) inasmuch as the equation can be rewritten in an equivalent manner as follows (see Gardner, Collier and Farr, 1934):

$$\psi = kp/\mu + \int_{z_0}^{z} k\rho g dz/\mu \tag{4.3.1.5a}$$

$$q = -\operatorname{grad} \psi \tag{4.3.1.5b}$$

Both representations by potentials are valid only if the integrals are univalent.

It will never be possible to distinguish between the two differential representations of Darcy's law if experiments are performed which use only constant-viscosity fluids and porous media of homogeneous permeability. The writer is not aware of any experiments reported in the literature which have employed porous media with a known change of permeability; therefore, a great uncertainty exists as to whether the differential form of Darcy's law given in equation (4.3.1.2) or that given in (4.3.1.3) is the correct one.

There have been attempts to justify one of the two forms upon general reasons. The possibility of the representations (4.3.1.4) or (4.3.1.5) hinges on the univalence of the integrals that occur in those formulas. The existence of this univalence is by no means *a priori* certain. If it could be shown, for example, for thermodynamic reasons, that the first of the integrals (4.3.1.4) is always univalent and the other (4.3.1.5) is not, this would be a strong indication that the first form is the correct representation of Darcy's law. It would *not prove* this, however, as the differential forms (4.3.1.2) and (4.3.1.3) are *always* possible and a decision cannot be made upon any other grounds than decisive experiments.

The univalence of the integral in (4.3.1.4a) is obvious from its definition provided that, for the fluid, the connection between p and ρ is univalent; i.e., provided that there is no hysteresis in the relation of pressure and density of the fluid. The existence of the potential ϕ will therefore be assured under quite general conditions. This is not the case for the existence of the velocity potential. No proof of either the existence or the non-existence of the velocity potential has come to the attention of the writer to date. Hubbert (1940) set out to show that the form (4.3.1.4a) would be the only correct one because of thermodynamic considerations. Allegedly, this fact is to follow from the second principle (i.e., the entropy principle) of thermodynamics, but Hubbert's proof does not seem to be convincing. Furthermore, even if it were shown that the velocity potential does not generally exist, the corresponding *differential* formulation (4.3.1.3) of Darcy's law would still be possible.

In general, the force potential form (4.3.1.2) of Darcy's law has been preferred to (4.3.1.3) in the literature because it is easier to treat analytically. There is another indication of its correctness, however, *viz.*, the fact that, if it is extended to multiple phase flow (see Part VIII of this book), it leads to the relative permeability concept. Relative permeability is actually variable through the porous medium during a flow experiment and therefore a distinction between equations

(4.3.1.2) and (4.3.1.3) is important. The fact that the relative permeability equations, as generalized from (4.3.1.2) and not from (4.3.1.3), lead to an adequate description of the experimental facts, is a strong indication for the correctness of the assumption of a force-potential.

Either differential form of Darcy's law, (4.3.1.2) or (4.3.1.3), is by itself not sufficient to determine the flow pattern in a porous medium for given boundary conditions as it contains three unknowns (q, p, ρ). Two further equations are therefore required for the complete specification of a problem. One is the connection between ρ and p of the fluid:

$$\rho = \rho(p), \qquad (4.3.1.6)$$

and the other a continuity condition, *viz.*:

$$- P \, \partial \rho / \partial t = \mathrm{div}\,(\rho q) \qquad (4.3.1.7)$$

where as usual P is the porosity and t time. With the help of these equations, one can eliminate all the unknowns except p which leads to either of the following equations, according to whether a force-potential (a) or a velocity potential (b) was assumed:

$$(a) \quad P \partial \rho / \partial t = \mathrm{div}\,[(\rho k / \mu)\,(\mathrm{grad}\,p - \rho g)] \qquad (4.3.1.8a)$$

$$(b) \quad P \partial \rho / \partial t = \mathrm{div}\,\{\rho[\mathrm{grad}\,(kp/\mu) - k\rho g / \mu]\} \qquad (4.3.1.8b)$$

where, again, g is a vector pointing *down*ward.

4.3.2. Compressible Porous Media. After a differential form of Darcy's law has been derived for fixed porous media, it must be generalized for compressible porous media.

Attempts at a description of the flow through compressible porous media have been made by Froehlich (1937) and Jacob (1940). A more or less consistent theory of the subject was given later in a series of papers by Shchelkachev (1946a, 1946b, 1946c; cf. also Leĭbenzon 1947, p. 84).

The theory of Shchelkachev is based upon the assumption that (a) both fluid and medium are elastic bodies following Hooke's law; i.e. (see secs. 1.6 and 2.2.1),

$$d\rho/dp = \rho \beta_f, \qquad (4.3.2.1)$$

$$dP/dp = \beta_m, \qquad (4.3.2.2)$$

where ρ is the density of the fluid and P the porosity of the solid, β_f and β_m are compressibility coefficients, assumed to be constant; and that (b) the permeability of the medium does not change during compression.

Accordingly, the whole effect of the compressibility of the medium is due to an alteration of the continuity equation. The latter can be shown to become:

$$\mathrm{div}\,(\rho q) = -\left(P + \frac{\beta_m}{\beta_f}\right) \frac{\partial \rho}{\partial t}. \qquad (4.3.2.3)$$

The theory of Shchelkachev thus simply replaces P by $(P + \beta_m/\beta_f)$ in the normal continuity equation, and therefore also in the usual differential formulation (4.3.1.8) of Darcy's law. It will be shown later that assumption of a finite compressibility of the medium is equivalent to assuming a "super-compressibility" of the liquid.

The theory of Shchelkachev is incomplete inasmuch as no dependence of the permeability on the prevailing fluid pressure has been assumed. We have mentioned above the investigations of Tiller (1953a, also 1953b, 1954) who postulated such a dependence (cf. equation 4.2.2.2). On the other hand, Tiller did not deduce the continuity equation corresponding to the dependence of porosity on fluid pressure as assumed by him (which is different from 4.3.2.2), but proceeded directly to integrate the Darcy equation (4.3.1.8) with a variable permeability according to (4.2.2.2). He was only interested in the constant-rate case applicable to certain filtration experiments and hence assumed q to be a constant. In order to make a consistent hydrodynamic theory out of Tiller's results, one would have to deduce the corresponding consistent continuity condition. This condition would be different from Shchelkachev's owing to the difference in the assumption of the dependence of P as a function of p. These investigations have not yet been brought to a close.

4.3.3. Anisotropic Porous Media.
Early investigations on the subject of anisotropic media have been made by Schaffernak (1933), Vreedenburgh and Stevens (1936), and Aravin (1937). The theory of flow through anisotropic media has been developed by Ferrandon (1948), Ghizetti (1949), and Litwiniszyn (1950). A further review has been given by Irmay (1951b). Litwiniszyn arrived at his theory by analogy with the process of diffusion, whereas Ferrandon gave an actual theoretical derivation of the formulas. Both theories are essentially identical.

The theory of Ferrandon (see Scheidegger, 1954) assumes that the contribution to the quantity q_n of flow through unit area in the direction $\boldsymbol{n}$ (components n_i), from elementary flow tubes parallel to the direction $\boldsymbol{m}$ (components m_i) whose combined cross-sectional area is equal to $cd\Omega$ ($d\Omega$ denoting the solid angle), is proportional to the gradient of a force-potential ϕ in the direction of m (cf. equation 4.3.1.4). Thus one has (the summation convention being applied):

$$dq_n = - kcn_i m_i \rho (\partial \phi / \partial x_j) m_j d\Omega / \mu.$$

Here, k and c are, of course, functions of m_i, such that one can set upon integration:

$$q_n = - n_i \rho \partial \phi / \partial x_j \int kcm_i m_j d\Omega / \mu = - n_i (k_{ij}/\mu) \rho (\partial \phi / \partial x_j)$$

where $k_{ij} = k_{ji}$. This can be written in vectorial form as follows:

$$q = - (\bar{k}/\mu)\rho \, \text{grad} \, \phi \qquad (4.3.3.1)$$

where $\bar{k}$ is a symmetric tensor consisting of the components k_{ij}. It can be properly referred to as the "permeability tensor" of the porous medium.

The fact that the permeability in anisotropic porous media has the form of a symmetric tensor, leads immediately to the following conclusions:

(i) In general the force potential gradient grad ϕ and filter velocity q are not parallel.

(ii) There are three orthogonal axes in space along which the force potential gradient and the velocity do have the same direction. These axes are termed the "principal axes" of the permeability tensor.

The task remains to relate the components k_{ij} of the permeability tensor to directional permeability measurements. Thus, let us investigate what happens in a narrow tube that was cut out at the direction n of an anisotropic porous medium. Obviously, the filter velocity q must be parallel to n; let it be denoted by q_n. The drop in ϕ along the tube denoted by ϕ_n is then given from equation (4.3.3.1) by

$$\phi_n = n \text{ grad } \phi = \mu n \overline{k^{-1}} n q_n / \rho.$$

If we define the "directional permeability" k_n' by the expression

$$k_n' = \mu q_n / (\phi_n \rho), \tag{4.3.3.2}$$

we obtain at once:

$$k_n' = 1/(n \overline{k^{-1}} n) \tag{4.3.3.3}$$

Let us now choose the principal axes of the permeability tensor as co-ordinates (corresponding permeabilities being k_1, k_2, k_3) and denote the angles of n with those axes by α, β, γ. Then (4.3.3.3) yields:

$$\frac{1}{k_n'} = \frac{\cos^2 \alpha}{k_1} + \frac{\cos^2 \beta}{k_2} + \frac{\cos^2 \gamma}{k_3}. \tag{4.3.3.4}$$

This is the central equation of an ellipsoid, if

$$r = \sqrt{k_n'} \tag{4.3.3.5}$$

is plotted on the corresponding directions of n. One has thus the following *theorem*:

If the square root of the directional permeability k_n' is measured on all of the corresponding directions in a point of an anisotropic porous medium, then one obtains an ellipsoid. The axes of the latter are in the direction of the principal axes of permeability, their length being equal to the square root of the principal permeabilities.

An *alternate way* to define directional permeability physically is by choosing a system in which the pressure drop is given by the boundary conditions, and by measuring that component of the velocity which is parallel to the pressure gradient. The directional permeability k_n'' is in this case defined by applying

Darcy's law formally to the given pressure gradient and the velocity component parallel to it. Thus, one has

$$k_n'' = n\bar{k}n \qquad (4.3.3.6)$$

as a second definition of "directional permeability."

Let us again choose the principal axes of the permeability tensor as coordinate axes (corresponding permeabilities being k_1, k_2, k_3) and denote the angle of n with those axes by α, β, γ. Then (4.3.3.6) yields

$$k_n'' = k_1 \cos^2 \alpha + k_2 \cos^2 \beta + k_3 \cos^2 \gamma. \qquad (4.3.3.7)$$

This is the central equation of an ellipsoid if

$$r = \frac{1}{\sqrt{k_n''}} \qquad (4.3.3.8)$$

is plotted on the corresponding directions of n. One has thus the following alternate *theorem*:

If the inverse square root of the directional permeability k_n'' is measured on all of the corresponding directions in a point of an anisotropic porous medium, then one obtains an ellipsoid. The axes of the latter are in the direction of the principal axes of permeability, their length being equal to the inverse square root of the principal permeabilities.

The fact that there are two different kinds of "directional permeability," denoted above by k_n' and by k_n'', is somewhat disconcerting. Whether one should consider k_n' or k_n'' in any particular case depends, of course, on the type of measurement that has been made. Fortunately, it is possible to show that in most cases the difference between the two types of directional permeability is quite negligible (Scheidegger, 1956).

In order to show this, let us consider the two-dimensional case. We then have

$$\frac{1}{k_n'} = \frac{\cos^2 \alpha}{k_1} + \frac{\sin^2 \alpha}{k_2}, \qquad (4.3.3.9)$$

$$k_n'' = k_1 \cos^2 \alpha + k_2 \sin^2 \alpha, \qquad (4.3.3.10)$$

and the ratio of the two types of directional permeability yields

$$k_n''/k_n' = \cos^4 \alpha + (k_1/k_2 + k_2/k_1) \cos^2 \alpha \sin^2 \alpha + \sin^4 \alpha$$

$$= 1 + \frac{(k_1 - k_2)^2}{k_1 k_2} \cos^2 \alpha \sin^2 \alpha. \qquad (4.3.3.11)$$

The right hand side of this expression reaches a maximum value for $\alpha = 45°$; it is then equal to:

$$\left.\frac{k_n''}{k_n'}\right|_{\max} = 1 + \frac{1}{4} \frac{(k_1 - k_2)^2}{k_1 k_2}. \qquad (4.3.3.12)$$

The maximum excess m over 1 of the ratio k_n''/k_n' (i.e., the maximum per cent error committed if k_n' is equalled to k_n'') is therefore given by

$$m = \frac{1}{4} \frac{(k_1 - k_2)^2}{k_1 k_2}. \qquad (4.3.3.13)$$

It is convenient to express this error m in terms of the excess n of the ratio k_1/k_2 over $1 (k_1 > k_2)$ which yields

$$m = \frac{1}{4} \frac{n^2}{1+n}. \qquad (4.3.3.14)$$

This shows that, for excesses n smaller than 1, the corresponding error m is proportional to n^2 and therefore negligible; for excesses n larger than 1, the corresponding error m goes asymptotically to $n/4$. It is therefore seen that k_n' and k_n'' are, in approximation, freely interchangeable. The relative error thereby committed increases with the excess of the ratio $k_1/k_2 (k_1 > k_2)$ over 1, but never exceeds 1/4 of this excess.

Physical measurements of directional permeability have been reported by Johnson and Hughes (1948, see section 4.2.2). Unfortunately, these authors plotted k_n as a polar permeability diagram instead of $1/k_n^{\frac{1}{2}}$. They therefore obtained shapes resembling somewhat a circle or also the figure "eight." Scheidegger (1954) recalculated the results of Johnson et al. and analysed their data in the light of the tensor theory of Ferrandon. These investigations yielded a substantiation of the tensor theory of permeability. An example of the fit of measured values of $1/k_n^{\frac{1}{2}}$ with an ellipse is shown in figure 12.

4.4. The Measurement of Permeability

The measurements of permeability can be performed using any of the forms of Darcy's law. In a homogeneous medium, the position of k/μ "inside" or "outside" the gradient is immaterial.

Thus, experiments are performed whereby in a certain system a pressure drop and a flow rate are measured. The solution of Darcy's law corresponding to the geometry of the system and to the fluid employed is calculated, and a comparison between the calculated and the experimentally found results immediately yields the only unknown quantity k. Usually horizontal linear systems are used as they are most easy to calculate. However, radial arrangements are also often used and even more complicated ones when the permeability of some medium is to be measured "in situ."

Physically, permeability measurements are very simple. They mainly involve questions of experimental technique. Methods have been discussed in general terms, for example by Kawakami (1933), Plummer et al. (1936), Manegold (1938), Koppuis and Holton (1938), and Eastman and Carlson (1940). A neat and simple

§4.4 THE MEASUREMENT OF PERMEABILITY 67

apparatus has been described by Pollard and Reichertz (1952); it is shown in figure 13. The drawing in the figure is self-explanatory.

Permeabilities in various substances may have a wide range. In table IV, we give a compilation of some representative values. As was the case in table I

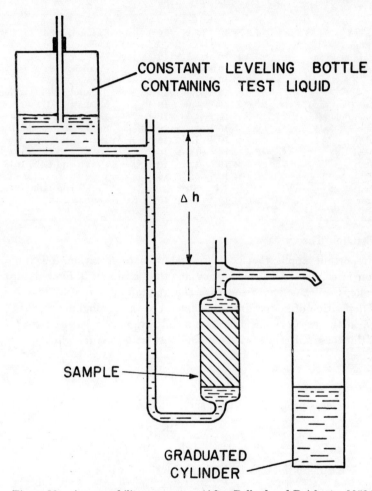

Figure 13. A permeability apparatus. (After Pollard and Reichertz, 1952.)

referring to porosities, "range" does not signify that the indicated values are the extreme limits of permeability which may be found in the respective substances; rather it signifies the range in which one is likely to find the permeability. As in table I, literature references are given indicating the provenance of the values shown.

TABLE IV
Representative Values of Permeability for Various Substances

Substance	Permeability range (Permeability in cm²)	Literature reference
Berl saddles	1.3×10^{-3} – 3.9×10^{-3}	Carman, 1938
Wire crimps	3.8×10^{-5} – 1.0×10^{-4}	Carman, 1938
Black slate powder	4.9×10^{-10} – 1.2×10^{-9}	Carman, 1938
Silica powder	1.3×10^{-10} – 5.1×10^{-10}	Carman, 1938
Sand (loose beds)	2.0×10^{-7} – 1.8×10^{-6}	Carman, 1938
Soils	2.9×10^{-9} – 1.4×10^{-7}	Aronovici and Donnan, 1946
Sandstone ("oil sand")	5.0×10^{-12} – 3.0×10^{-8}	Muskat, 1937
Limestone, dolomite	2.0×10^{-11} – 4.5×10^{-10}	Locke and Bliss, 1950
Brick	4.8×10^{-11} – 2.2×10^{-9}	Stull and Johnson, 1940
Leather	9.5×10^{-10} – 1.2×10^{-9}	Mitton, 1945
Cork board	3.3×10^{-6} – 1.5×10^{-5}	Brown and Bolt, 1942
Hair felt	8.3×10^{-6} – 1.2×10^{-5}	Brown and Bolt, 1948
Fibre glass	2.4×10^{-7} – 5.1×10^{-7}	Wiggins et al., 1939
Cigarette	1.1×10^{-5}	Brown and Bolt, 1942
Agar-Agar	2.0×10^{-10} – 4.4×10^{-9}	Pallmann and Deuel, 1945

4.5. Filtration Theory

An important application of Darcy's law is to filtration. The process of filtration is conducted in such a manner that a slurry is filtered through a filter cake, the latter increasing with the amount of material deposited. The rate of deposition of material is proportional to the throughput of filtrate.

Thus, in the case of constant pressure filtration, one has at time t a filter cake of thickness h. Darcy's law, therefore, yields for the throughput q (scalarly):

$$q = \frac{k}{\mu} \frac{p_2 - p_1}{h}$$

However, dh/dt is proportional to q, such that one has

$$\frac{dh}{dt} = cq = -\frac{k}{\mu} \frac{p_2 - p_1}{q^2} \frac{dq}{dt}. \tag{4.5.1}$$

The last is a differential equation whose solution for q is

$$q = \sqrt{\frac{k(p_2 - p_1)}{2\mu(ct + c')}}.$$

The total volume V of filtrate filtered through a unit area from the beginning of the process of filtration is therefore:

$$V(t) = \int_{t_0}^{t} q\, dt = (1/c)\sqrt{\tfrac{1}{2}(k/\mu)(p_2 - p_1)(ct + c')} + V_0.$$

This can be rewritten as follows, by including everything that is constant into the new constants G, K, and t_0

$$(V + G)^2 = K(t + t_0). \tag{4.5.2}$$

This is the form in which the fundamental filtration equation has been given by Ruth, Montillon and Montonna (1933). These authors also checked their equation by a great number of experimental data.

The process of filtration has been investigated since by many authors; but refinements which were thus achieved belong in a treatise on filtration rather than in a treatise on flow through porous media. Some of these investigations, if they are concerned with the physics of the flow through porous media, will be discussed later in their proper context.

PART V

SOLUTIONS OF DARCY'S LAW

5.1. General Remarks

Solutions of hydrodynamic problems concerning the flow of homogeneous fluids in porous media are obtained by solving the equation (4.3.1.8) for particular boundary conditions. Most solutions to be discussed are based upon the first form of Darcy's law (4.3.1.8a). Unless the contrary is indicated, it is therefore to be assumed that (4.3.1.8a) is preferred to (4.3.1.8b).

The differential equation in question is very closely related to the equations of diffusion and to that of heat conduction. Thus, solutions obtained with the intention of solving the heat-conduction equation can often be taken over directly for the hydromechanics in porous media.

The treatment of the differential equation resulting from Darcy's law is, actually, a discipline of mathematics and has very little to do with the physics of the problem. General methods that are applicable may be found in appropriate mathematical texts such as that by Courant and Hilbert (1943). Much mathematical material which is directly applicable to the flow in porous media has also been accumulated by Carslaw and Jaeger (1947) in connection with their study of heat conduction. In this present chapter, we shall not give a comprehensive review of the methods that are applicable to the solution of the differential equation; that belongs in a study on the theory of functions and mathematical analysis. Rather, we shall give a review of physical conditions for which solutions have been achieved. Thus, solutions which have been obtained with regard to problems other than the flow of fluids through porous media (such as heat conduction problems) will not be found here although they might have direct analogues bearing upon the present study.

The physical conditions of flow for which solutions might be sought are (i) steady state flow, (ii) gravity flow with a free surface, and (iii) unsteady state flow. Of these, steady state flow solutions for incompressible fluids are most easily obtained; they are simply represented by solutions of the Laplace equation. Except for a few other special cases, Darcy's law, however, leads to nonlinear differential equations. The analytical methods to deal with these are quite involved and lengthy so that efforts have been directed towards scaling phenomena and also towards the experimental representation by analogous effects.

A comprehensive review of solutions of Darcy's law may be found in the recent book by Polubarinova-Kochina (1952) which has already been mentioned in the Introduction. Some solutions have also been accumulated in the book by Muskat (1937b) mentioned earlier.

5.2. Steady State Flow

5.2.1. Analytical Solutions.

The steady state is characterized by the vanishing of the partial time derivatives of the physical quantities such as the density, velocity, etc. Equation (4.3.1.8) thus reduces to:

$$\text{div}\,[(\rho k/\mu)\,(\text{grad}\,p - \rho g)] = 0 \qquad (5.2.1.1)$$

(note that g is a vector pointing *down*ward, but co-ordinates increase *up*ward). If, furthermore, the fluid is incompressible and the porous medium homogeneous, one has:

$$\text{lap}\,p = 0, \qquad (5.2.1.2)$$

where "lap" denotes the Laplace operator. This is the well-known differential equation of Laplace applicable in many instances in physics, and the general methods to which this differential equation is amenable lead to valid solutions in the present case. Thus, if two-dimensional problems be considered, methods based upon the theory of functions of complex variables are applicable as well as the technique of using Green's function. A review of such methods is given in the books of Muskat (1937) and of Polubarinova-Kochina (1952). Golubeva (1950) discussed the utility of curvilinear co-ordinates in two-dimensional motion, and Shaw and Southwell (1941) and Dykstra and Parsons (1951) applied relaxation methods to the same problem. Other general methods for the treatment of steady-state flow have been discussed by Pavlovskiĭ (1936c), Rossbach (1936–8), Mandel (1939), Meleshchenko (1940), Rizenkampf (1940), Kalinin (1944), Polubarinova-Kochina and Falkovich (1947), Vinogradov and Kufarev (1948), and Charniĭ (1950a). Furthermore, Kozlov (1940a) and Polubarinova-Kochina (1940b) showed how steady-state solutions for anisotropic flow can be obtained by a simple co-ordinate transformation onto principal axes.

As a classical example of a steady-state solution we give the solution for two-dimensional radial flow of an incompressible fluid into a well which is completely penetrating the fluid-bearing medium. Assuming that the well is a cylinder of radius R_0, with pressure there p_0, and that the pressure at distance R_1 from the well is p_1, it is easy to verify that the required solution is

$$Q = \frac{2\pi k}{\mu \log\text{nat}\,(R_1/R_0)}\,(p_1 - p_0), \qquad (5.2.1.2a)$$

where Q is the total discharge per unit time and unit penetration length of the well; k is the permeability of the medium and μ is the viscosity of the fluid.

If the fluid in motion is compressible, then equation (5.2.1.1) does not reduce to the Laplace equation. However, it has been pointed out by Leĭbenzon (1947, p.132) that in the case of gases it does reduce to the Laplace equation if the following substitution is made

$$\chi = \int_{p_0}^{p} \rho\, dp \qquad (5.2.1.3)$$

which leads to

$$\text{lap } \chi - \text{div}(\rho^2 \mathbf{g}) = 0.$$

In the case of gases, the second term is much smaller than the first and can therefore be neglected. This also holds with horizontal flow of compressible liquids. One thus obtains

$$\text{lap } \chi = 0. \qquad (5.2.1.4)$$

The study of the steady flow of compressible fluids in most instances is thus reduced to the discussion of the same differential equation as that encountered in the study of the steady flow of incompressible fluids. One has, for instance, in the linear case (co-ordinate x) for an ideal gas ($cp = \rho$)

$$\chi = (c/2)p^2 = c' + bx.$$

With the boundary conditions $p(x = 0) = p_0$ and $p(x = L) = p_L$ one has

$$p^2 = p_0^2 + x(p_L^2 - p_0^2)/L. \qquad (5.2.1.5)$$

If now Darcy's law is expressed at $x = 0$, one obtains

$$q_0 = -(k/\mu)(dp/dx)_0 = -(k/\mu)(p_L^2 - p_0^2)/(2p_0 L). \qquad (5.2.1.6)$$

This equation is commonly used in permeability determination if a gas is used as percolating fluid. A similar formula can be deduced for a radial arrangement.

The steady flow of liquids in simple geometrical patterns has been analysed by O'Brien and Putnam (1941); by Collins (1952), who discussed some problems of horizontal flow; by Kazarnovskaya and Polubarinova-Kochina (1943), who studied seepage-flow through non-horizontal layers; by Polubarinova-Kochina (1950b), who investigated the effects of sources and sinks on a surface; by Kalinin (1952), who studied seepage through a wedge consisting of two different materials; and by Scheidegger (1953), who investigated the flow through cavities in porous media. Vlasov (1951) and Lelli (1952) studied the flow in inclined layers, and Hubbert (1940) and Pavlov (1942), the effect of a layer of altered permeability.

There are a great number of papers seeking steady state solutions of Darcy's law with respect to certain applications. One of the most important applications is the study of the flow from strata into wells. Muskat (1937) and Polubarinova-Kochina (1952) discuss a series of such solutions in their books. Furthermore, Shchelkachev and Pȳkhachev (1939) investigated the interference of wells and the theory of water-drive in oil fields; Polubarinova-Kochina (1942) discussed some general aspects of the flow into wells; Muskat (1943) studied the effect of casing perforations on well productivity and Pȳkhachev (1944) gave a further general discussion of flow into wells. Dodson and Cardwell (1945) investigated wells completed with slotted liners. Girinskiĭ (1946a) gave a generalization of some cases for wells in rather complex natural conditions, Myatiev (1946) discussed the flow to a well in heterogeneous ground, and Segal (1946) the flow into wells from the standpoint of potential theory. In addition, Polubarinova-Kochina (1947) gave the hydraulic theory in a stratified medium, Lukomskaya (1947) studied the radial flow of liquids into wells, and Kufarev (1948, 1950a) gave the solution of Darcy's law for flow into wells from a circular boundary. Similarly, Luthin and Scott (1952) gave a numerical analysis of flow through aquifers towards wells in difficult conditions such as variable permeability.

Related to the study of the steady flow into wells in the problem of drainage of water-saturated strata by drainage tubes, etc., Kirkham (1940, 1945, 1949, 1950a, 1950b) has written a number of papers bearing upon that subject. Similar problems have also been investigated by Meleshchenko (1936), Topolyans'kiĭ (1940, 1947), Vcrigin (1949a), Luthin and Gaskell (1950), Citrini (1951), Boreli (1953), and Day and Luthin (1954).

Much work has also been done on the study of seepage underneath engineering structures, especially dams (with and without sheet-piling). Again, Muskat (1937) and Polubarinova-Kochina (1952) gave reviews of a wide variety of solutions that have been obtained. Such problems have also been investigated by Meleshchenko (1936b), Girinskiĭ (1937, 1941), Kozlov (1939, 1940a, 1940b, 1941), Aravin (1940), Bil'dyug (1940), Feylessoufi (1940), Verigin (1940), Rossbach (1942), Segal (1942), Segal and Gantmakher (1942), Kalinin (1943), Nedriga (1937), Sauvage-de-St. Marc (1947), Verigin (1947), Mansur and Perret (1948), Voshchinin (1948), Terracini (1949, 1950), McNown and Hsu (1950), Fil'chakov (1951), Mandel (1951), Shima (1951), Kochina and Polubarinova-Kochina (1952), Sokolov (1952) and by Guevel (1953).

The details of these investigations are mainly concerned with particular engineering requirements and are of only limited interest in relation to fundamental physical concepts. The reader is therefore referred to the original papers and to the monograph of Polubarinova-Kochina (1952) for further study. As an illustrative example, one solution only is presented here in some detail.

Following Day and Luthin (1954), we consider the ideal case of a level, uniform porous medium of depth h underlain by a completely permeable substratum (see figure 14). In practice the porous medium may be a sandy soil and the substratum a loosely packed gravel bed. Numerous long furrows extend parallel to one another across the surface of the medium and are separated by a distance a between centres. They are semi-circular in cross-section and have a radius equal to r. The furrows are filled with fluid and are kept full by continuous additions (problem of irrigation of a field). It will be permissible to assume that a steady state is reached after fluid has flowed continuously into the substratum for a considerable time.

The pattern of flow, which is two-dimensional, can be deduced straightforwardly. In a medium of homogeneous permeability, the formulation using the velocity potential may be used and one has

$$\psi = \frac{k}{\mu}(p + \rho g z), \tag{5.2.1.7}$$

$$q = -\operatorname{grad} \psi, \tag{5.2.1.8}$$

$$\operatorname{lap} \psi = 0. \tag{5.2.1.9}$$

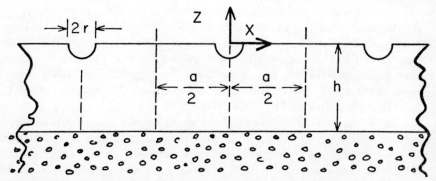

Figure 14. Geometrical arrangement corresponding to Day and Luthin's (1954) solution of a steady-state seepage problem. (After Day and Luthin, 1954.)

The boundary conditions can be satisfied using the method of images (Kirkham, 1949), and one obtains

$$\psi = A \sum_{m=-\infty}^{m=+\infty} \log \operatorname{nat} \left\{ \frac{\cosh\left[\pi(x-ma)/(2h)\right] - \cos\left[\pi z/(2h)\right]}{\cosh\left[\pi(x-ma)/(2h)\right] + \cos\left[\pi z/(2h)\right]} \right\} + \psi_0 \tag{5.2.1.10}$$

where ψ_0 is the velocity potential at the top of the substratum; m is an integer having successive values running from $-\infty$ to $+\infty$; and A is a constant.

As is well known from the theory of the Laplace equation, one can introduce a stream function φ, connected with ψ by the Cauchy-Riemann differential equations

$$\partial \psi / \partial z = \partial \varphi / \partial x; \; \partial \psi / \partial x = -\partial \varphi / \partial z \tag{5.2.1.11}$$

so that the lines $\varphi = $ constant represent streamlines. It is easy to verify that the expression for φ, corresponding to (5.2.1.11), is:

$$\varphi = 2A \sum_{m=-\infty}^{m=+\infty} \tan^{-1} \left\{ \frac{\sinh\left[\pi(x-ma)/(2h)\right]}{\sin\left[\pi z/(2h)\right]} \right\} + \varphi_0 \tag{5.2.1.12}$$

Thus, the flownet can be plotted. A solution conforming to some particular constants, is shown in figure 15 (after Day and Luthin, 1954).

5.2.2. Solutions by Analogies. The Laplace equation (5.2.1.2) occurs in many contexts in physics. Therefore, solutions of that equation can be obtained by

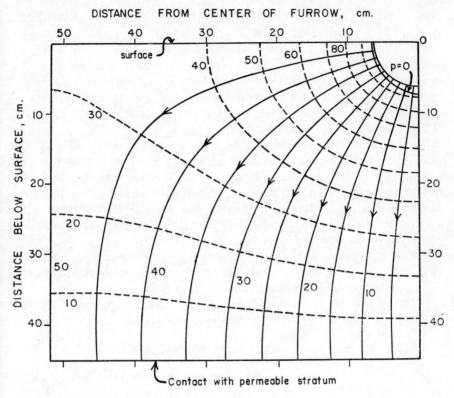

Figure 15. Streamlines in a particular seepage problem. (After Day and Luthin, 1954.)

performing suitable experiments which are themselves governed by the Laplace equation. In this manner, it is often possible to set up an analogue to a certain problem of steady flow through porous media and thus to avoid the tedious job of solving the Laplace equation analytically. We shall discuss some of the possible analogies (Scheidegger, 1953).

(a) *The flow of electricity.* The steady flow of electricity in a conducting medium is governed by Laplaces's equation:

$$\text{lap } \Phi = 0 \qquad (5.2.2.1)$$

$$i = - \text{ grad } \Phi \qquad (5.2.2.2)$$

where $E = \Phi/\rho$ is the electric potential, ρ the conductivity, and i the current density. Furthermore, on the boundary of two media one of which is very much more conducting than the other, the potential will be constant. This gives a means of duplicating various boundary conditions of flow through porous media. Thus, in a model experiment one would use a trough of some electrolyte of conductivity ρ and insert electrodes, corresponding to the boundary conditions. In such an experiment ρ, E, i will be proportional to k/μ, p, q, respectively, if care is taken that corresponding units are used. In an actual experiment, one would also have to contend with polarization effects at the electrodes; it would therefore be advisable to use alternating current of moderately high frequency. Experiments of this type for the study of flow through porous media have been successfully completed by Ram, Vaidhianathan and Taylor (1935), Lee (1948), and Dolcetta (1948).

(b) *The flow of heat.* The flow of heat in a heat-conducting medium is also governed by Laplace's equation; unfortunately it is not generally feasible to use this fact for constructing experimental analogies of the flow through porous matter, as it is quite difficult to measure heat flow accurately. However, many analytical solutions of Laplace's equation have been developed in connection with the study of heat flow (Carslaw and Jaeger, 1947), and these can be at once applied to porous flow problems.

(c) *The distribution of stresses.* It has been suggested that it might be possible to duplicate the flow lines by stress lines in stressed materials and to examine the latter by means of experimental stress analysis.

Indeed, a plane stress state could be used to represent two-dimensional flow. The trace of the stress tensor $\mathscr{T}$ fulfils Laplace's equation in two dimensions:

$$\text{lap Trace } \mathscr{T} = \text{lap } (S_1 + S_2) = 0. \tag{5.2.2.3}$$

Here, S_1 and S_2 are the stress components in any two orthogonal directions in the point under consideration, S_3 is zero because of the assumption of a plane stress state.

Unfortunately, it is in general quite difficult to realize experimentally boundary conditions which would correspond to a flow-pattern. The proposition can be better realized if one restricts oneself to two dimensions; thus a soap film may be used to represent two-dimensional flow. The differential equation for a stretched membrane is of the form

$$\text{lap } z(x, y) = 0 \tag{5.2.2.4}$$

where z is the lateral ordinate as a function of x and y, two arbitrarily chosen Cartesian co-ordinates normal to z. Then, the curves of equal height would correspond to equipotential lines in the flow. Such experiments have actually been conducted as part of the study of heat flow (Wilson and Miles, 1950). In connection with flow through porous media, the analogy has been discussed by Hansen (1952a).

(d) *The flow of viscous fluids.* Under certain circumstances, the Navier-Stokes equation (2.2.2.1), too, reduces to a Laplace equation. It is therefore possible to use the phenomenon of viscous flow for a representation of steady flow in porous media. This method has been applied by Günther (1940) and Barron (1948).

(e) *Mechanical scaling.* It is possible finally to represent a large-scale flow phenomenon in porous media by a small-scale one. All that is necessary is to make the geometrical dimensions to scale. The pressures may be scaled in any desirable way since the Laplacian of (ap) is zero if that of p vanishes. A series of experiments using such scaling has been discussed in the book of Muskat (1937). In addition, the method has also been applied by Plummer and Woodward (1936) and by Kirkham (1939).

5.3. Gravity Flow with a Free Surface

5.3.1. Physical Aspects of the Phenomenon.
A peculiar special case of the determination of steady state flow with gravity occurs where a fluid has a "free" surface. In fact, this is a problem of multiple-phase flow; above the "free surface" there will be another fluid. Nevertheless, if that other fluid be a gas, and the original fluid a liquid, one can disregard the motion of the gas with respect to that of the liquid and assume that the gas pressure is constant in its entire domain. Thus one can express the assumption that there is a free interface between the two fluids by the condition that on this free interface the liquid pressure is uniform and that any streamline having one point in common with this free surface lies *entirely* within it. In truth, no such free interfaces actually exist as there will always be a finite region within which the liquid saturation drops from 1 to zero. Nevertheless, we shall see later that sharp saturation discontinuities can exist such that the assumption of an actual "free surface" may well be sensible.

A further peculiar phenomenon may occur if the notion of a free surface is accepted. Should this free surface intersect an "open" boundary of the porous medium, i.e., a boundary which is "open" to the gas and upon which therefore the pressure is uniform and equal to that of the gas, then liquid will seep out from that boundary *below* the intersection with the free surface. Such a boundary is termed "surface of seepage." The analytical condition for a surface of seepage is that the pressure in the fluid be constant and equal to that at the free surface. In contrast to the free surface, however, a surface of seepage need not contain any streamlines. The physical picture of the free seepage has been analysed on many occasions, for instance, by Hamel (1934), Vedernikov (1947a), Emery and Foster (1948), and Laurent (1949).

Thus, the analytical conditions of the flow with a free surface are fully determined. To summarize, one has to find a solution of the Laplace equation (5.2.1.2) with the boundary conditions such that there is a "free surface" which contains streamlines and on which the fluid pressure is constant. On "open boundaries" below the intersection with the free surface, the liquid pressure must be constant and equal to that on the free surface. On impermeable boundaries, of course, the condition is, as usual, that the normal component of the filter velocity vanish.

Needless to say, it is extremely difficult and tedious to find analytical solutions conforming to the above boundary conditions. Breitenöder (1952), Muskat (1937) and, especially, Polubarinova-Kochina (1952) reviewed some methods that have been applied successfully, notably one employing hodograph transformations. In a medium of homogeneous permeability, the formulation using the velocity potential may be used:

$$\psi = (k/\mu)(p + \rho g z), \qquad (5.3.1.1)$$

$$\mathbf{q} = -\operatorname{grad} \psi, \qquad (5.3.1.2)$$

$$\operatorname{lap} \psi = 0. \qquad (5.3.1.3)$$

In two dimensions (x horizontal, y vertical co-ordinates), this leads to the well-known possibility of representing everything in terms of complex numbers with $z = x + iy$. The hodograph transformation is characterized by setting

$$u = \partial \psi/\partial x \qquad (5.3.1.4)$$

$$v = \partial \psi/\partial y \qquad (5.3.1.5)$$

which again can be represented in a complex plane.

The hodograph method has the advantage that the free surfaces, whose shape is unknown in the original formulation of the problem, are determined in the hodograph plane: they are simply circles with known parameters. In addition, the surfaces of seepage are also determined prior to an actual analytical solution of the problem. Thus, the "floating" boundary conditions of the original problem become fixed boundary conditions after a hodograph transformation has been made.

The use of hodograph transformations in connection with problems of seepage flow has been developed chiefly by Hamel (1934). Further discussions of analytical attempts to solve the problem may be found in the books of Breitenöder (1942), Muskat (1937) and Polubarinova-Kochina (1952). The latter author especially gives a large collection of methods that are applicable to the problem. She has also written many original papers on the subject (Polubarinova-Kochina 1938, 1939a, 1939b, 1939c, 1939d, 1940a, 1941). Other methods have been discussed by Zhukovskiĭ (1923), Vedernikov (1948), Davison (1936), Muskat (1936), Rizenkampf (1938a, 1938b), Vibert (1939), Rossbach (1941), Gersevanov (1943), Kalinin (1944), Girinskiĭ (1946b), and Malcor

(1948, 1950). A further review may also be found in a paper by Polubarinova-Kochina and Fal'kovich (1947).

Owing to the difficulties of obtaining analytical solutions of the problem, graphical methods have been tried (Nahrgang, 1954). In the case of radial flow (vanishing azimuthal component of the velocity), the flow equation, using the notation of velocity potential (cf. 5.3.1.1), reduces to:

$$\frac{\partial^2 \psi}{\partial r^2} + \frac{1}{r}\frac{\partial \psi}{\partial r} + \frac{\partial^2 \psi}{\partial z^2} = 0. \tag{5.3.1.6}$$

It is well known from the theory of the Laplace equation that one can introduce a stream function φ, defined by the differential equations:

$$\partial \varphi / \partial z = 2\pi r \partial \psi / \partial r \tag{5.3.1.7}$$

$$-\partial \varphi / \partial r = 2\pi r \partial \psi / \partial z. \tag{5.3.1.8}$$

The stream function φ, in turn, satisfies the following differential equation:

$$\frac{\partial^2 \varphi}{\partial r^2} - \frac{1}{r}\frac{\partial \varphi}{\partial r} + \frac{\partial^2 \varphi}{\partial z^2} = 0, \tag{5.3.1.9}$$

as can easily be verified. The lines ψ = const. represent in the r–z plane equipotential curves, and the lines φ = const. represent streamlines. The streamlines and the equipotential lines form an orthogonal net of curves. Therefore, from equation (5.3.1.7–8) one can show that the differential of the potential can be expressed in terms of the differential of the stream function as follows:

$$d\varphi = 2\pi r \frac{dn}{ds} d\psi \tag{5.3.1.10}$$

where dn is the differential along a potential line and ds is the differential along a streamline.

Using differences (denoted by Δ), instead of differentials, and postulating

$$\Delta \varphi = \Delta \psi, \tag{5.3.1.11}$$

we have:

$$\Delta s = 2\pi r \Delta n. \tag{5.3.1.12}$$

This initiates the possibility of a graphical solution which is particularly suited to the gravity flow problem. We have to construct a net of orthogonal curves satisfying (5.3.1.12) and the boundary conditions. This can be done by the method of trial and error (cf. p. 82 and figure 16).

Again, because of the difficulties involved in obtaining rigorous solutions conforming to the assumptions basic to the present considerations, approximate procedures have been developed. The best known is that by Dupuit (1863) as modified by Forchheimer (cf., e.g., 1930). If it is assumed that for small inclinations of the free surface of a gravity flow system the streamlines may be taken as horizontal, and, furthermore, that the corresponding velocities are proportional to the slope of the free surface, one readily arrives at the following differential equation for the height h of the free surface above the (horizontal) impermeable bed of the system (see Muskat, 1937, p. 360):

$$\operatorname{lap}(h^2) = 0. \tag{5.3.1.13}$$

Unfortunately, the assumptions basic to this simple theory do not seem to be quite warranted so that the entire theory is severely questioned (see Muskat, 1937, p. 359). The assumptions of the Dupuit-Forchheimer theory have also been used by Boussinesq (1904) to construct a theory of the time variations of the free surface. The latter author arrived thus at the following equation (see, for example, Leïbenzon, 1947):

$$P\partial h/\partial t = (k\rho g/\mu) \operatorname{div}(h \operatorname{grad} h) \tag{5.3.1.14}$$

where the symbols have their usual meaning. This equation can be simplified for certain cases so as to become linear. However, since it is based on the Dupuit-Forchheimer assumptions and therefore is subject to the same criticisms, a further discussion is hardly warranted.

It has been observed that the flux formulas obtained from the Dupuit-Forchheimer theory give much better results than the underlying assumptions might lead one to expect. Muskat (1937) showed that one can arrive at the Dupuit-Forchheimer *flux* formulas by another approximate theory which is free of the Dupuit assumptions. The Dupuit-Forchheimer theory has been discussed more recently by Leïbenzon (1947) and Reinius (1947). In spite of its basic shortcomings, it is still widely used for engineering calculations.

The analytical difficulties of obtaining solutions of the gravity-flow problems have also prompted people to try experimental methods. Among these methods, electrical analogues seemed the most hopeful. However, the free surface cannot be duplicated in electrical models as there are no corresponding boundaries possible for electrical currents. Thus, the procedure is one of trial and error, of shaping the electrical model in such a manner that part of its boundaries corresponds to a free surface. Muskat (1937) discussed in his book (p. 318) the difficulties involved. Other investigations have been made by Wyckoff and Reed (1935) and by Fil'chakov (1949).

5.3.2. Specific Solutions. The principles outlined above in connection with the description of free surfaces and seepage surfaces have been applied to specific problems. Again, such specific applications are nothing but exercises in mathematics and not of too much interest as far as the physical aspects of flow through porous media are concerned. Muskat (1937) reviewed a number of such solutions in his book. A further review of such applications has been given in a paper by Rizenkampf (1938a, 1938b), by Polubarinova-Kochina and Fal'kovich (1947), and especially in the recent book by Polubarinova-Kochina (1952).

Turning now to original investigations, it was probably the paper of Hopf and Trefftz (1921) that started the whole issue of gravity-flow. Since this paper, a great variety of problems have been studied by many authors, e.g., Lefranc

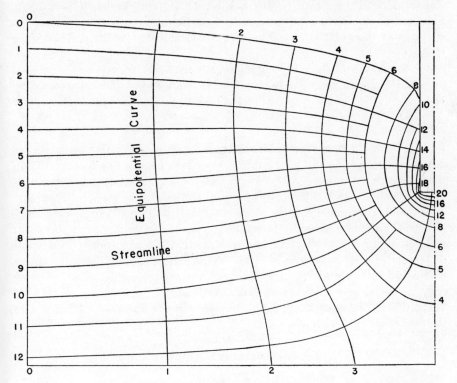

Figure 16. Flow lines and equipotential curves in a partially penetrating well. (After Nahrgang, 1954.)

(1938), Vibert (1938a, 1938b), Davison and Rosenhead (1940), Rossbach (1941), Ivakin (1947), Reinius (1947), Chwalla (1948), Merkel et al. (1949), Verigin (1949b, 1949c, 1950), Uchida (1950) and Irmay (1953). In particular, the seepage through earthen dams from a reservoir has received much attention. Such problems have been treated by Pavlovskiĭ (1936a), Davison (1936b), Nel'son-Skornyakov (1937a, 1937b, 1937c, 1940a, 1940b, 1941a, 1941b, 1947), Girinskiĭ (1938, 1939), Numerov (1939, 1940a, 1940b, 1942, 1946), Voshchinin (1939), Polubarinova-Kochina (1940c), Vedernikov (1945, 1947b), Fandeev (1947), Uginchus (1947), Mikhaĭlov (1951, 1953) and Kalinin (1952). In connection with studies of agricultural development (drainage and irrigation of land), the problem of drainage into and seepage from channels, tubes, etc., has been profusively investigated, *viz.*, by Pavlovskiĭ (1936a, 1936b), Aravin (1936, 1937b), Khomovskaya (1937), Nel'son-Skornyakov (1940d, 1941c), Bazanov (1938, 1940), Vedernikov (1939b, 1940), Numerov (1940c, 1940d, 1940e), Rizenkampf (1940b), Rizenkampf and Kalinin (1941), Childs (1947), Van Deemter (1949), Engelund (1951), Nasberg (1951), Sokolov (1951) and Verigin (1951, 1953). Finally, free surface flow (of water) into wells has also received some attention in papers by Vibert (1939), Boulton (1942, 1951), Hansen (1952b), and Nahrgang (1954).

For the details of these investigations, the reader is referred to the original papers. As an example, we only show here in figure 16 the solution obtained by Nahrgang (1954) by his graphical method for a partially penetrating well.

5.4. Unsteady State Flow

5.4.1. Unsteady State Flow in Linear Approximation.

If one is to consider the general patterns of flow in porous media, one has to insert ρ as a function of p in the differential equation (4.3.1.8). In general, the latter thereupon becomes nonlinear, which makes it difficult to treat it analytically.

However, it is possible to linearize the differential equation in certain approximations of two cases: (i) if the fluid is a liquid of constant compressibility; and (ii) if the porous medium is elastic. Both these cases lead to identical differential equations, as pointed out by Shchelkachev (1946a, 1946b, 1946c), so that assuming a certain compressibility of the porous medium has the same effect as assuming a corresponding compressibility of the fluid.

The differential equation (4.3.1.8b) becomes in the linear approximation, in the case of a liquid with

$$\rho = \rho_0 \exp\{\beta_f(p - p_0)\}, \qquad (5.4.1.1)$$

according to (2.2.1.1), and upon neglect of the gravity term:

$$P\beta_f \frac{\partial \rho}{\partial t} = \frac{k}{\mu} \operatorname{lap} \rho. \qquad (5.4.1.2)$$

It should be noted, however, that the equation does *not* become linear if the gravity term is not neglected.

Similarly, it has been shown that assuming an elastic porous medium alters the continuity equation to read (4.3.2.3). If, therefore, this be inserted into Darcy's law and the same approximations made as above, one obtains

$$(P\beta_f + \beta_m)\frac{\partial \rho}{\partial t} = \frac{k}{\mu} \operatorname{lap} \rho. \qquad (5.4.1.3)$$

It is seen that (5.4.1.2) and (5.4.1.3) are identical if the terms $P\beta_f$ and $P\beta_f + \beta_m$ are identical. Thus, assuming an unrealistic "super"-compressibility of the fluid may well account for a neglected compressibility of the porous medium.

It is often possible to neglect powers of β_f higher than the first in a Taylor expansion of (5.4.1.1):

$$\rho \simeq \rho_0 + \beta_f \rho_0 (p - p_0). \tag{5.4.1.4}$$

Using this approximation, it becomes obvious that equation (5.4.1.3) can be written as follows:

$$(P\beta_f + \beta_m) \frac{\partial p}{\partial t} = \frac{k}{\mu} \operatorname{lap} p. \tag{5.4.1.5}$$

This shows that, under the assumptions outlined above, the same linear differential equation applies to the pressure as to the density.

As mentioned above, if gravity terms are kept in Darcy's law, the differential equations do not become linear in ρ. If the equations are to be linearized, some other approximations have to be made. Thus, it has been observed by Werner (1946a, 1946b) that for small compressibilities the continuity equation for the fluid can, in analogy with (5.4.1.4), be written approximately as follows:

$$\operatorname{div} \mathbf{q} = -\bar{\rho}(P\beta_f + \beta_m) \frac{\partial \phi}{\partial t}, \tag{5.4.1.6}$$

with $\bar{\rho}$ denoting the average fluid density in the medium and ϕ the usual force potential (see 4.3.1.4). Combined with Darcy's law (4.3.1.4b), one thus obtains (neglecting second order terms in β):

$$\operatorname{lap} \phi = \frac{\mu}{k} (P\beta_f + \beta_m) \frac{\partial \phi}{\partial t}, \tag{5.4.1.7}$$

which is linear.

Thus, in approximation, the same equations hold for the density ρ, the pressure p, and the force potential ϕ.

Equations (5.4.1.2) (5.4.1.3) and (5.4.1.7) are of the form of the heat conductivity equation. Thus, the general methods applicable for solving problems of heat conduction are also applicable to the transient flow in porous media. We shall not discuss here the analytical methods that may be used to treat (5.4.1.1), etc.; such methods may be found in treatises on mathematical methods (e.g., Courant and Hilbert, 1943) and in Carslaw and Jaeger's (1947) book on heat conduction. We shall list, however, a series of papers containing solutions of (5.4.1.1) with particular reference to problems of transient flow in porous media.

Muskat (1937) gives a review of such solutions up to about 1936. A review of a great number of more modern solutions has also been given by Polubarinova-Kochina and Fal'kovich (1947), as well as in the recent book of Polubarinova-Kochina (1952).

Many solutions are concerned with the flow into wells and drains, and connected effects. Papers dealing with such solutions have been written by

Hurst (1934), Theis (1935), Galin (1945, 1951a), Polubarinova-Kochina (1945), Cooper and Jacob (1946), Kalinin (1948), Charniĭ (1949), Kufarev (1948, 1950a, 1950b), Miller, Dyes and Hutchinson (1950a, 1950b), Shchelkachev (1951), Horner (1941), Jacob and Lohmann (1952), and Hurst (1953). Van Everdingen and Hurst (1949) discussed the application of the Laplace transformation to such well problems; their work was reviewed later by Chatas (1953a, 1953b, 1953c). Furthermore, Brownscombe and Kern (1951) developed a graphical method applicable to two-dimensional problems.

The effect of seepage from reservoirs through porous strata has been investigated by Jacob (1946), Polubarinova-Kochina (1940a, 1950b), and Verigin (1940d). The large-scale behaviour of fluid masses in porous media has been investigated in connection with a study of the runoff-curves of aquifers by Werner (1946a, 1946b), by Werner and Sundquist (1951), by Kochina (1951), by Polubarinova-Kochina (1951b), and by Pilatovskiĭ (1953a, 1953b). In addition, Werner and Norén (1951) studied the possibility of progressive waves in aquifers.

Owing to limitations of space, it is again impossible to give a detailed outline of all the methods that are applicable to the solution of the linearized non-steady state equation, However, to demonstrate at least one representative case, a solution is given here in somewhat more detail.

We take the case of the linearized equation for a compressible porous medium in accordance with the theory of Shchelkachev (1946b). Considering the simplest, i.e., linear arrangement, we can write the fundamental equation as follows (5.4.1.5):

$$a^2 \frac{\partial^2 p}{\partial x^2} = \frac{\partial p}{\partial t} \qquad (5.4.1.8)$$

with

$$a^2 = \frac{k}{\mu(P\beta_f + \beta_m)}. \qquad (5.4.1.9)$$

Particular boundary conditions can be stated as follows:

$$p = p_0 \text{ for } x = 0 \text{ at all times}$$
$$p = p_1 \text{ for } x = L \text{ at all times} \qquad (5.4.1.10)$$
$$p = p_0 \text{ everywhere for } t = 0.$$

The solution of this problem can be shown to be

$$p = p_0 - (p_0 - p_1)\frac{x}{L} + \frac{2}{\pi}(p_0 - p_1)\sum_{n=1}^{\infty}\frac{(-1)^{n-1}}{n} e^{-\frac{n^2\pi^2 a^2}{L^2}t} \sin\left(\frac{n\pi}{L}x\right) \qquad (5.4.1.11)$$

as one may verify easily by differentiation.

In order to give a series of "standard" solutions, it is convenient to introduce the following dimensionless parameters:

$$\frac{x}{L} = \varepsilon; \quad 2\frac{a^2}{L^2}t = F; \quad \frac{p_0 - p}{\varepsilon(p_0 - p_1)} = W. \qquad (5.4.1.12)$$

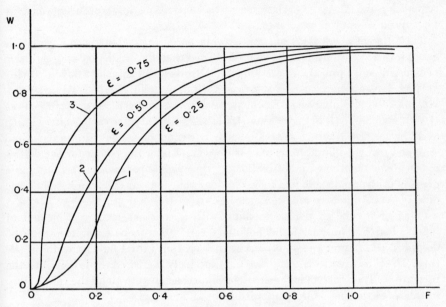

Figure 17. A series of solutions for linear non-steady state flow in porous media. (After Shchelkachev, 1946.)

In this notation, the solution of our problem becomes

$$W = 1 - \frac{2}{\varepsilon\pi}\sum_{n=1}^{\infty}\frac{(-1)^{n-1}}{n}\exp\left(-\frac{n^2\pi^2 F}{2}\right)\sin(n\pi\varepsilon). \qquad (5.4.1.13)$$

The solutions corresponding to this expression can then be plotted as a series of curves as shown in figure 17.

5.4.2. General Unsteady State Flow. If the approximations considered in the last paragraphs are not made, the unsteady state equation becomes nonlinear. Owing to analytical difficulties, only relatively few solutions of this case are available.

One of the most important cases is the study of transient gas flow through porous media. In this case, the gravity term can be neglected; however, the Darcy equation is nevertheless nonlinear even if the simplest connection possible (i.e., $\rho = cp$ corresponding to an ideal gas) between pressure p and density ρ of a gas is used.

Methods applicable in such cases have been discussed by Muskat and Botset (1931), Khristianovich (1941), Hetherington, MacRoberts and Huntington (1942), Leïbenzon (1945), Polubarinova-Kochina (1948), MacRoberts (1949), Kalinin (1950), Piskunov (1951), Aronofsky and Jenkins (1952), Green and Wilts (1952), Roberts (1952), Bruce, Peaceman Rachford and Rice (1953), and Aronofsky and Ferris (1954). Some of these solutions have been obtained by high-speed computing machines.

The differential equation of transient flow becomes nonlinear also if gravity is taken into account, even if the simplified (and therefore inadequate) description of Boussinesq is applied (see section 5.3.1). Problems of transient flow involving gravity have been treated by Devison (1937), Galin (1945, 1951a, 1951b), Kalinin and Polubarinova-Kochina (1947), Charniĭ (1950b), Evdokimova (1950), Baumann (1951), Kochina (1951, 1953), Polubarinova-Kochina (1951), and Werner (1953); the last treatment is based upon the Boussinesq equation.

Again, owing to limitations of space, we must refer the reader for the details of the above-mentioned investigations to the original papers. As an illustration of some of the applicable methods, only one solution can be given here in some detail (after Aronofsky and Jenkins, 1952):

Consider a tube of porous medium with a constant cross section and of infinite length. The initial and boundary conditions to be considered may be described as follows: Let p_0 represent the constant initial fluid pressure in the tube. The pressure at one end, $x = 0$, is suddenly increased (or lowered) to the pressure p_1. The problem is then to determine the flow rate along the tube at any instant of time. The fluid is supposed to be an ideal gas.

The problem is solved if it is possible to obtain, subject to the boundary conditions, a solution of the differential equation

$$\frac{\partial^2 (p/p_0)^2}{\partial x^2} = \frac{2P\mu}{p_0 k} \frac{\partial}{\partial t} \left(\frac{p}{p_0}\right) \tag{5.4.2.1}$$

which follows from (4.3.1.8) for an ideal gas in the one-dimensional case. Equation (5.4.2.1), being non-linear, cannot easily be treated analytically. Therefore, it appears one should use a numerical approximation procedure. In order to use such a numerical procedure, one replaces (5.4.2.1) by the following difference equation:

$$\left(\frac{p}{p_0}\right)_{x,\,t+\Delta t} = \frac{\Delta t}{(\Delta x)^2 a} \left[\left(\frac{p}{p_0}\right)^2_{x+\Delta x,\,t} + \left(\frac{p}{p_0}\right)^2_{x-\Delta x,\,t} - 2\left(\frac{p}{p_0}\right)^2_{x,\,t}\right] + \left(\frac{p}{p_0}\right)_{x,\,t}, \tag{5.4.2.2}$$

where $(p/p_0)_{x,\,t}$ equals a dimensionless pressure at distance x and time t; $(p/p_0)_{x+\Delta x,\,t}$ equals the same dimensionless pressure at distance $x + \Delta x$ and time t; $(p/p_0)_{x,\,t+\Delta t}$ equals the dimensionless pressure at distance x and time $t + \Delta t$, and a equals $2P\mu/(p_0 k)$.

In order to ensure stability of equation (5.4.2.2) one has to restrict Δt such that

$$\Delta t \leq \frac{a}{4}(\Delta x)^2. \qquad (5.4.2.3)$$

Substituting the maximum Δt permitted by equation (5.4.2.3) in (5.4.2.2) yields the difference equation in the following usable form

$$p_{x,t+\Delta t} = \tfrac{1}{4}[p^2{}_{x+\Delta x, t} + p^2{}_{x-\Delta x, t} - 2p^2{}_{x,t}] + p_{x,t} \qquad (5.4.2.4)$$

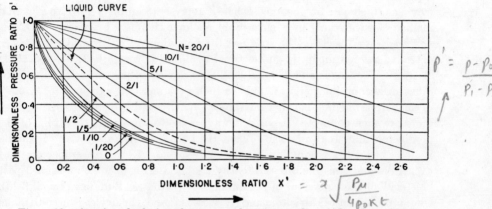

Figure 18. A series of solutions for non-steady state gas flow. (After Aronofsky and Jenkins, 1952.)

where p_0 is unity. The difference equation can now be solved with ease for a given set of initial and boundary conditions, except that the computations are tedious and lengthy. A high-speed (electronic) computing machine should therefore be used.

Applying the above procedure to the case under consideration, Aronofsky and Jenkins (1952) obtained the results presented in figure 18. The ordinate in this figure is a dimensionless pressure p'

$$p' = (p - p_0)/(p_1 - p_0) \qquad (5.4.2.5)$$

which is the ratio of pressure change at any point within the porous medium to the pressure change at the face, $x = 0$. The abscissa is the dimensionless parameter x'

$$x' = x\sqrt{\frac{P\mu}{4p_0 kt}} \qquad (5.4.2.6)$$

which brings all the data into a single plot. It is necessary to use a family of curves to present the results as the ratio N

$$N = p_1/p_0 \qquad (5.4.2.7)$$

is allowed to take on different values.

It may be remarked that the solution represented in figure 18, although it was developed for an infinitely long tube, is also valid for a finite tube if one is interested only in events happening before a substantial pressure change occurs at the opposite end of the tube. Many physical problems arise which meet this requirement.

5.4.3. Experimental Solutions. The analytical difficulties met in dealing with the differential equations of section 5.4 are quite considerable even in the linearized case. This is especially true if the boundaries of the porous medium are irregular in shape: for instance, when it represents an oil field. Experimental methods for the solutions of such problems have therefore been developed.

A tool often used for the study of the behaviour of compressible liquids in extended porous media is an electrical analogue (Bruce, 1943; Morris, 1951; Patterson, Montague and Wiess, 1951). Such devices approximate the linearized differential equation. They are built in such a manner that a domain (or "zone") of the porous medium is represented by a capacitor and a resistor in series. Such units are then connected in various ways so as to represent the extent of the porous medium. It can be shown that the behaviour of an array of such units is described by an equation approximate to (5.4.1.1). The proper sizes of capacitors and resistors are selected by trial and error to match a known past "history" of the behaviour of the fluid in the porous medium (e.g., an oil field). Once a proper set of capacitors and resistors is found, the device is used to make predictions for the future behaviour. Unfortunately, a past history can be matched in many ways which, in turn, may give somewhat different "predictions," so that there is the possibility of occasional prediction errors. In a similar fashion, Green and Wilts (1952) developed an electrical network analogue whereby the non-linear gas equations can be solved by successive approximation.

PART VI

PHYSICAL ASPECTS OF PERMEABILITY

6.1. Empirical Correlations

The concept of permeability as introduced in the earlier parts of this review permits a phenomenological description of the flow through porous media in a certain velocity domain. However, an actual understanding of the phenomena can be obtained only if the concept of permeability can be reduced to more fundamental physical principles.

It seems intuitively clear that the property of "permeability" should be linked with other properties of a porous medium, *viz.*, capillary pressure curves, internal surface area, etc., since all such properties are the manifestation of the geometrical arrangement of the pores. However, to uncover such relationships will be possible only if one is able to understand exactly how all these properties are conditioned by the geometrical properties of the pore system.

A direct approach to finding relationships between the various properties of porous media is by an attempt to establish empirical correlations. A most obviously sought after correlation will be that between *porosity and permeability*. A simple consideration of theoretical possibilities of the structures of porous media, however, makes one realize that a general correlation between porosity and permeability cannot exist. It is obviously quite possible for two porous media of the same porosity to have entirely different permeabilities. Thus if a correlation function between the two quantities is sought after, it cannot be unique. Therefore, most empirical correlations contain some other factors, usually vaguely identified with alleged geometrical quantities. They are, however, nothing but undetermined factors used in order to make the data fit the desired equations. There are even a series of claims for "general" relationships, usually supposed to be true for "average" porous media, whatever that means. In the early attempts (up to about 1933) of the oil industry to elucidate the behaviour of reservoirs, no distinction was even made between porosity and permeability, implying proportionality between the two properties.

Some of the claims for empirical relationships between porosity and permeability are listed by Jacob (1946b), Franzini (1951), and Hudson and Roberts (1952). Thus, Mavis and Wilsey (1936) claim to have found that k is proportional to P^6 or P^5, Büche (1937) gives a similar relationship and Rose (1945) sets k proportional to P^n (with n indeterminate). Other equations have been proposed by Brevdy (1948) and by Shuster (1952). Most of these proportionalities contain further factors when written as equations and stipulate that, in order to be applicable, these "other factors" must be kept constant. Allegedly, these "other factors" are representative of certain geometrical properties of the porous medium. The factor used mostly contains the "grain diameter" δ in some form (usually to the power two) which, in turn, is thought equivalent to the "pore diameter" (whatever this is). As neither of these quantities usually is properly defined, their introduction and proper "adjustment" means nothing more or less than the introduction of an undetermined constant which is adjusted from case to case so as to make the "correlations" properly valid. The introduction of these quantities would be significant only if there were an independent means of measuring them.

The influence of porosity upon permeability has been discussed in more general terms (i.e., without giving a specific equation, but by supplying curves or qualitative discussions) by Baver (1940) and by Bulnes and Fitting (1945). A further discussion is due to Cloud (1941) who definitely finds that there is no sensible relation between porosity and permeability.

It is therefore obvious that no simple correlation between porosity and permeability can exist. The next more involved correlation that might be looked for is between *structure and permeability*. Again, of course, the meaning of "pore structure" is somewhat arbitrary. One may understand that this means the "pore size distribution," as obtained from capillary pressure curves. If such correlations were found, one would have, in fact, obtained a correlation between capillary pressure curve and permeability; whether this is also a correlation between pore size distribution and permeability depends on whether one wants to consider capillary pressure as representative of pore size distribution, a problem which has been discussed in Part III of this study. The correlations between capillary pressure and permeability are mainly based upon theoretical considerations which will be discussed later. The dependence of permeability on other "structural" parameters of the pore space has been discussed, upon an experimental basis, by Nelson and Baver (1940), O'Neal (1940), and Backer (1951). None of these investigations seem to show very promising results.

For unconsolidated porous media, correlations between *permeability and grain size distribution* have been attempted. Such correlations have been reported by Tickell (1935), Mavis and Wilsey (1937), Prockat (1940), Krumbein and Monk (1942), Sen-Gupta and Nyun (1943), and Pillsbury (1950). Most of these analyses were made on similar types of materials and therefore it was possible for the individual authors to claim such correlations. However, it seems hardly credible that general correlations should exist, i.e., that Raschig rings and pebbles that happen to be screened by the same set of sieves should have the same permeability. It has thus been pointed out by Makhl (1939) that a comparison of products as to the coefficient of permeability is possible only if the tests are made under similar conditions and the products are of the same type. Cloud (1941) found that there is no correlation and Griffiths (1952a) observed that, in oil sands, the grain size distribution is correlated with certain mineral compositions rather than with permeability. It so happens that specific minerals often form beds of certain characteristic permeabilities owing to their usual mode of breaking up and packing. In this manner, a correlation between grain size distribution and permeability might actually be obtained, but certainly not one that is expressible in mathematical terms; it is rather the case that certain minerals normally show both a characteristic grain-size distribution and a characteristic permeability.

In view of the fact that a simple correlation between grain size distribution (as obtained from sieve analysis) and permeability does not seem to exist, an attempt has been made to introduce further parameters. Thus, the influence of a parameter indicative of the "angularity" or "roundness" of the grains on permeability has been investigated by Tickell et al. (1933), Tickell and Hiatt (1938), and the influence of the "packing" of the grains has been investigated by Martin et al. (1951).

Attempts have further been made to correlate the orientation of the grains with permeability. Such studies have been made by Griffiths (1950a, 1952b), Griffiths and Rosenfeld (1953), Martin et al. (1951), and Heiss and Coull (1952). The outcome of these studies seems to be that orientation has a definite effect, although the results cannot be represented by a simple correlation. Nevertheless, this may explain why beds of the same materials, packed to the same porosity, occasionally have different permeabilities.

There have also been attempts to find experimental correlations between properties of porous media other than those mentioned above. Thus Bjerrum and Manegold (1927), Buchanan and Heymann (1948), Thornton (1949), and Goring and Mason (1950) discussed the question of relationship between electrical and mechanical flow path in porous media and Kolb (1937) made an attempt to correlate water imperviousness with characteristics of adsorption and sand surface of mortars and concrete.

6.2. Capillaric Models

6.2.1. The Concept of Models.
The empirical attempts to establish correlations between various dynamical properties of porous media all seem to be futile unless certain additional parameters are introduced. Theoretical considerations, however, might be able to attach a physical significance to those parameters.

Such theoretical considerations will be based upon an analysis of the microscopic properties of flow. The flow through porous media, presumably, takes place along flow channels with local (pore) velocity v. The pore velocity, on the whole, must be larger than the filter velocity q, owing to the reduced space available for the particles to flow, as compared with the bulk volume of the porous medium with regard to which the filter velocity is calculated. A commonly accepted hypothesis for the connection between pore velocity and filter velocity is the following, known as the Dupuit-Forchheimer assumption:

$$v = \frac{q}{P}. \tag{6.2.1.1}$$

However, it should be noted that "pore velocity" has really no properly defined meaning, because the actual velocity of the fluid must be expected to fluctuate grossly within one flow channel and from flow channel to flow channel. The Dupuit-Forchheimer assumption *defines* an "average" pore velocity which may, or more often may not, be identical with what a particular microscopic flow theory would like to regard as "pore velocity." It should therefore be noted that the Dupuit-Forchheimer assumption cannot be regarded as basic, unless it is shown that the "pore velocity" under consideration *is* an actual statistical average over all the local velocities, as implied by that very assumption.

The simplest way to try to establish correlations theoretically is by representing the porous media by theoretical models which can be treated mathematically. Only trial and error can show which models exhibit the characteristic phenomena taking place in the porous medium and which do not. If a proper model is found, it can be substituted for an actual porous medium and one will be able to predict by calculation how the medium will behave under yet untried conditions. Relationships deduced from such models would be generally valid.

The simplest models that can be constructed are those consisting of capillaries. Scheidegger (1953) gave a review of such models. Previously, capillaric models had been discussed by Kawakami (1932), Smith (1932), Rainard (1947), Henderson (1949), Kustov (1949), Purcell (1949), Burdine, Gournay and Reichertz (1950), and Calhoun (1953). These models aim at correlating the permeability with either an "average" pore size or with the pore size distribution (i.e., capillary pressure) curve.

6.2.2. Straight Capillaric Model. The simplest capillaric model of the linear case is one representing a porous medium by a bundle of straight parallel capillaries (see, for example, Scheidegger, 1953) of uniform diameter $\bar{\delta}$. The flow Q through a capillary is then given by the well-known law of Hagen-Poiseuille (see equation 2.2.2.2).

$$Q = -\frac{\pi \bar{\delta}^4}{128\mu} \frac{dp}{dx}. \tag{6.2.2.1}$$

Here, μ is as usual the viscosity and dp/dx the pressure gradient along the capillary. If there are n such capillaries per unit area of cross section of the model, the flow per unit area q (or, the macroscopic or "filter" velocity) will be:

$$q = -\frac{n\pi \bar{\delta}^4}{128\mu} \frac{dp}{dx}. \tag{6.2.2.2}$$

As the flow can also be expressed by Darcy's law

$$q = -\frac{k}{\mu}\frac{dp}{dx}, \qquad (6.2.2.3)$$

it follows that

$$k = \frac{n\pi\bar{\delta}^4}{128}. \qquad (6.2.2.4)$$

The pore volume of the model (assuming unit cross-sectional area) is equal to $\tfrac{1}{4}n\pi x\bar{\delta}^2$, the length being denoted by x; thus the porosity is

$$P = \tfrac{1}{4}n\pi\bar{\delta}^2. \qquad (6.2.2.5)$$

Eliminating n from equations (6.2.2.4) and (6.2.2.5) yields the following equation:

$$k = \frac{P\bar{\delta}^2}{32}. \qquad (6.2.2.6)$$

If this equation is applied to an actual porous medium, $\bar{\delta}$ is a sort of "average" pore diameter.

It is known that this equation does not correctly represent the connection between permeability and porosity in porous media as it is actually observed. Therefore, the factor 32 is commonly replaced by some arbitrary factor T^2 where T is termed "tortuosity" or the like. Similarly, instead of $\bar{\delta}^2$ the "average specific surface area" S can be introduced. The specific internal area of the model (i.e., the ratio of area of the capillaries to volume of the model) is given by

$$S = n\pi\bar{\delta}.$$

Eliminating n by means of equation (6.2.2.5), one obtains

$$S = \frac{4P}{\bar{\delta}}.$$

Introducing this into the expression for the permeability, and taking all numerical factors together in the "tortuosity" T, the equation (6.2.2.6) takes the following form

$$k = \frac{P^3}{T^2 S^2}. \qquad (6.2.2.7)$$

A similar equation had been proposed long ago by Krüger (1918) on experimental grounds; later a theoretical justification was given by Kozeny (1927a), using, however, different reasoning altogether as will be shown later. Equations containing S, similar to (6.2.2.7), are now commonly called "Kozeny Equations."

The present procedure of adding arbitrary factors does not help one to understand the phenomena in the porous medium and it is therefore necessary to investigate what alterations would have to be made on the simple model of parallel capillaries to get a more adequate description of the flow phenomena.

6.2.3. Parallel Type Models. It is natural to stick first as closely as possible to the simple model of parallel capillaries treated in section 6.2.2.

First, one notices that this model gives a permeability in one direction only. All capillaries being parallel, there can be no flow orthogonal to the capillaries. The first modification of the relationship between porosity and permeability would therefore have to consist in putting one-third of the capillaries in each of three spacial dimensions. The permeability is thus lowered by a factor 3 and equation (6.2.2.6) will read

$$k = \frac{P\bar{\delta}^2}{96}. \tag{6.2.3.1}$$

The inclusion of the factor 3, of course, obliterates the Dupuit-Forchheimer assumption of $v = q/P$ which is thus not valid in the present models.

Apparently, this is not much better than the original equation. One of the main difficulties is that the average diameter $\bar{\delta}$ of the pores of an actual porous medium does not have much meaning. It will be desirable to reduce $\bar{\delta}$ to some value calculated from the "pore size distribution" $f(\delta)$ or from the "differential pore size distribution" $\alpha(\delta)$ as defined in Part I. In fact, this means aiming at a connection between capillary pressure and permeability since the pore size distribution is usually measured (more or less accurately) by the capillary pressure method (see Part III). Thus, we shall construct a model where all the capillaries permitting flow in a given direction are parallel to that direction, but vary in pore diameter, leading from one face of the porous medium through to the other. We shall call such a model a "parallel type model." The pore size distribution of the model and an actual porous medium will be made identical. A model of this type was devised by Purcell (1949) in a somewhat different fashion. The following exposition is taken from an earlier review by the writer (Scheidegger, 1953).

For a given pressure gradient equal to dp/dx, the mean velocity v of a flowing fluid through a capillary of radius δ is given by the Hagen-Poiseuille equation (6.2.2.1):

$$\frac{dp}{dx} = -32\frac{\mu v}{\delta^2}. \tag{6.2.3.2}$$

Now, the volume taken up by capillaries of pore diameter between δ and $\delta + d\delta$ parallel to the x-direction in a piece of the model of unit frontal area and thickness Δx is $\frac{1}{3}P\alpha(\delta)d\delta\Delta x$. (The factor $\frac{1}{3}$ is expressing that only one-third of the capillaries in the model are in one given direction, as explained above.) They have a frontal area of $\frac{1}{3}P\alpha(\delta)d\delta$. Thus the total quantity of fluid flowing through a unit area per unit time is

$$q = \tfrac{1}{3}P\int_0^\infty v\alpha(\delta)d\delta. \tag{6.2.3.3}$$

Inserting v from (6.2.3.2) yields

$$q = -\frac{dp}{dx}\frac{P}{96\mu}\int_0^\infty \delta^2\alpha(\delta)d\delta \qquad (6.2.3.4)$$

and by comparison with Darcy's law:

$$k = \frac{P}{96}\int_0^\infty \delta^2\alpha(\delta)d\delta. \qquad (6.2.3.5)$$

Thus, we obtain almost the same expression as equation (6.2.3.1), except that the average pore radius $\bar\delta$ has now a more exactly defined meaning. It is given by the equation:

$$\bar\delta^2 = \int_0^\infty \delta^2\alpha(\delta)d\delta. \qquad (6.2.3.6)$$

No fundamental deviation from equations (2.2.3.1) and (6.2.2.6) is therefore obtained. Particularly, the specific surface area and a tortuosity factor can again be introduced, and thus one finally arrives again at a "Kozeny" equation of the type of (6.2.2.7). It seems that parallel type models have not much advantage over the simple model of parallel capillaries; the essential consequences are identical.

6.2.4. Serial Type Models. A serious drawback of the parallel type models is that all the pores are supposed to go from one face of the porous medium right through to the other. This supposition is evidently far remote from what happens in an actual porous medium.

The opposite extreme picture of the one-dimensional case would be obtained by assuming that all the pore space is serially lined up, so that each particle of fluid would have to enter at one pinhole at one side of a porous medium and travel through very tortuous channels through all the pores and then emerge at one only pinhole at the other face of the porous medium. Obviously, this picture is just as unreal as that treated in section 6.2.3 and a realistic model lies somewhere in between the extremes. We shall refer to such models as "serial type models," as capillaries of different pore diameter are put together in series one after another (Scheidegger, 1953).

We shall therefore assume a model of length x where there are n capillaries per unit area in each dimensional direction of a pore diameter $\bar\delta$ and of length s. This model is assumed to represent a cylindrical porous medium of length x, porosity P, and "average" pore diameter $\bar\delta$. In this instance we shall not specify more clearly what is meant by "average" pore diameter, but think of it in rather indefinite terms as in section 6.2.2, where a simple model was discussed. Later on, we shall try to find a more clearly defined expression in terms of the function $\alpha(\delta)$ of the porous medium.

A model of the above type is a "serial type model," since the capillaries effecting the flow in any one direction may have contortions; in other words, s may be greater than x. Later on we shall vary the diameter of each channel along its length, but now we simply assume them all to be of the same "average" diameter $\bar{\delta}$.

If we assume a pressure drop equal to $p_2 - p_1$ over the length of the above model (one-dimensional case), the mean flow velocity v through each capillary is given by the Hagen-Poiseuille equation (6.2.3.2), neglecting capillary pressure differentials:

$$\frac{p_2 - p_1}{x} = -32 \frac{\mu v}{\bar{\delta}^2} \frac{s}{x}. \tag{6.2.4.1}$$

Thus the total flow per unit area and unit time through the model, if there are n capillaries per unit area, is:

$$q = n \frac{\pi}{4} \bar{\delta}^2 v = -\frac{n\pi}{128} \frac{x}{s} \frac{\bar{\delta}^4}{\mu} \frac{p_2 - p_1}{x}. \tag{6.2.4.2}$$

The porosity of the model is (making allowance for the fact that *each* spacial dimension has n capillaries per unit area, by introducing the factor 3)

$$P = \frac{3}{4} n\pi \bar{\delta}^2 \frac{s}{x}. \tag{6.2.4.3}$$

Eliminating n from (6.2.4.2) and (6.2.4.3) yields

$$q = -\frac{1}{96} \left(\frac{x}{s}\right)^2 \frac{\bar{\delta}^2}{\mu} P \frac{p_2 - p_1}{x}. \tag{6.2.4.4}$$

Comparison of this with Darcy's law yields

$$k = \frac{1}{96} \frac{P\bar{\delta}^2}{T^2} \tag{6.2.4.5}$$

if we set

$$ds/dx = s/x = T. \tag{6.2.4.6}$$

Again, this is of the form of equation (6.2.2.6). The quantity T plays the role of the arbitrary factor which had to be introduced into the equation (6.2.2.6) and which was termed there "tortuosity factor." Actually, $T = s/x$ would be an excellent measure of the tortuosity as it gives the ratio of the length of the flow channel for a fluid particle with respect to the length of the porous medium, if the model satisfactorily represents a porous medium.

However, it still remains that the average pore diameter $\bar{\delta}$ has no clearly defined meaning. It is therefore necessary to modify the above model by varying the diameter of each capillary along its length according to the pore size distribution function $\alpha(\delta)$ of the medium which the model is supposed to represent. In the porous medium, the fraction $\alpha(\delta)d\delta$ of the pore space has a pore diameter between δ and $\delta + d\delta$. Thus, the length ds, over which each capillary of the model has to be made of diameter between δ and $\delta + d\delta$, is given by the condition that the ratio of the capillary volume along ds with respect to the whole capillary volume must be equal to the ratio of pore space between δ and $\delta + d\delta$ with respect to the whole pore volume:

$$\frac{\dfrac{\pi}{4}\delta^2 ds}{\displaystyle\int_0^s \frac{\pi}{4}\delta^2 ds} = \alpha(\delta)d\delta. \qquad (6.2.4.7)$$

However, the integral is obviously

$$\int_0^s \frac{\pi}{4}\delta^2 ds = \frac{1}{3}\frac{Px}{n}, \qquad (6.2.4.8)$$

where x is again the length over which the pressure drop exists and n the number of capillaries per unit area. We obtain therefore:

$$\frac{\pi}{4}\delta^2 ds \equiv \frac{\pi}{4}\delta^2 T dx = \alpha(\delta)d\delta \frac{1}{3}\frac{Px}{n}. \qquad (6.2.4.9)$$

Furthermore, the continuity equation yields for the fluid velocities:

$$\frac{\pi}{4}n\delta^2 v = q. \qquad (6.2.4.10)$$

We assume now that the Hagen-Poiseuille equation (6.2.3.2) is valid for each infinitesimal length of the flow channels. This is, of course, an oversimplification of the facts since complicated effects will occur when a flow channel either contracts or expands. Nevertheless, this assumption is probably no worse than the assumption of a capillaric model in the first place. Thus, inserting (6.2.4.8–10) into the Hagen-Poiseuille equation, the following expression is obtained:

$$p_2 - p_1 = -\int \frac{32\mu v}{\delta^2} ds = -\int \frac{32\mu v}{\delta^2} T dx \qquad (6.2.4.11)$$

$$= \frac{512}{3\pi^2}\frac{\mu Px}{n^2} q \int_0^\infty \frac{\alpha(\delta)}{\delta^6} d\delta. \qquad (6.2.4.12)$$

Comparison with Darcy's law yields

$$\frac{1}{k} = \frac{512}{3\pi^2}\frac{P}{n^2}\int_0^\infty \frac{\alpha(\delta)}{\delta^6} d\delta. \qquad (6.2.4.13)$$

This expression for the permeability seems to be quite different from anything obtained above. It appears as if the permeability would decrease for increasing porosity. However, this is not the case. It should be observed that n, the number of capillaries per unit area, has a peculiar influence: If the porosity is to be increased with n being held constant, then this can be done only by lengthening the flow channels, and therefore the permeability will be less.

It is therefore better to introduce the tortuosity T as defined in (6.2.4.6) instead of n. It follows from (6.2.4.8):

$$\frac{\pi}{4} n\delta^2 = \frac{P}{3T}. \tag{6.2.4.14}$$

Multiplying this by $\alpha(\delta)d\delta$ and integrating,

$$\frac{P}{3T} = \frac{\pi}{4} n \int \delta^2 \alpha(\delta) d\delta. \tag{6.2.4.15}$$

Thus, substituting n from this expression into equation (6.2.4.13) yields

$$\frac{1}{k} = \frac{96 T^2}{P} \left[\int \delta^2 \alpha(\delta) d\delta \right]^2 \int \frac{\alpha(\delta)}{\delta^6} d\delta. \tag{6.2.4.16}$$

This is identical with what was obtained above by assuming a more simplified model (equation (6.2.4.5)), if one sets

$$\frac{1}{\bar{\delta}^2} = \left[\int \delta^2 \alpha(\delta) d\delta \right]^2 \int \frac{\alpha(\delta)}{\delta^6} d\delta. \tag{6.2.4.17}$$

The "average pore radius" $\bar{\delta}$ introduced above has thus again been given a more specific meaning as expressed by the last equation (6.2.4.17).

6.2.5. Comparison with Tests. It remains to compare the performance of the models discussed above with actual tests on porous media. The first model of parallel capillaries has very little meaning, since the average pore diameter $\bar{\delta}$ is not properly defined. It is of course obvious that, if $\bar{\delta}$ is *defined* by equation (6.2.2.6), any porous medium will suit the model. But then $\bar{\delta}$ has very little to do with an average pore diameter as obtained from pore size distribution and may at best be introduced as a "Kozeny constant" or something similar.

A proper modification of the model of parallel capillaries which states plainly the connection of k and P with the pore size distribution, is expressed by equation (6.2.3.5). In order to compare this equation with actual porous media, one needs the function $\alpha(\delta)$ for the latter. This function may be obtained from pore size distribution analyses, for example by the mercury injection method (see section 3.4.3). Then, one can calculate the value of the integral

$$\int_0^\infty \alpha(\delta) \delta^2 d\delta$$

and compare it with the expression $96k/P$ obtained from permeability and porosity data. At this point, one difficulty becomes apparent. This is that the integral is most affected by the value of $\alpha(\delta)$ for large pore sizes—and those cannot be measured by the mercury injection technique. If the integral is carried well into the region of large pores or vugs, its magnitude becomes out of all proportion and many powers of ten too large for what it should be, namely, equal to $96k/P$.

One can try to save the situation by saying that the vugs would have to be treated separately as enclosed cavities affecting the permeability, as discussed earlier (in section 5.2). Then it would have to be assumed that the integral must be cut off at some universal cut-off diameter; but an investigation shows that the cut-off diameter varies at least within a factor 100. This observation is substantiated by Henderson (1949) who shows that the permeability calculation from pore size distribution based on the model of parallel capillaries does *not* work.

To come to the serial type models, it is at once apparent that the integrals in the final formula (6.2.4.16) are very sensitive to errors in $\alpha(\delta)$ for very low and very high values of the argument δ. These are just the values of δ for which the pore size distribution function is very difficult to measure. On the other hand, there is an additional parameter T (the tortuosity) so that there is no doubt that the model fits any porous medium, simply by adjusting T correctly.

6.3. Hydraulic Radius Theories

6.3.1. Principles of Theory.
From the above description of the capillaric model theories it is apparent that more elaborate models must be used in order to obtain a satisfactory understanding of the hydromechanics of fluids in porous media.

A set of theories, thus, is based upon the assumption that a porous medium is equivalent to a series of channels; the latter are, however, assumed to be somewhat more elaborate than in the capillaric models. These theories all make use of the fundamental observation that the permeability, in absolute units, has the dimension of an area or of a length squared. It may be argued, therefore, that a length should be characteristic for the permeability of a porous medium. Such a length may be termed "hydraulic radius" of the porous medium, and is presumably linked with the hypothetical channels to which the porous medium is thought to be equivalent. A possible measure of a hydraulic radius would, for instance, be the ratio of volume to surface of the pore space.

Thus, the basic concept of the hydraulic radius theories is a consequence of dimensional considerations. One has to add, therefore, that the permeability, in addition to being proportional to some hypothetical square length, may also be dependent on any dimensionless quantities, notably on some function of the porosity (as the latter is dimensionless). Therefore the hydraulic radius theories assume the following basic expression for the permeability:

$$k = c \frac{m^2}{F(P)} \qquad (6.3.1.1)$$

where m is the "hydraulic radius," $F(P)$ is the "porosity factor," and c is some dimensionless constant which could be incorporated into the porosity factor.

The problem is, then, to find a physical significance for m and $F(P)$. General discussions of such questions as were outlined above may be found in the works of Terzaghi (1925), Nutting (1927), Nemenyi (1933), Manegold (1937, 1938), Manegold and Solf (1937, 1938), Jacob (1946a), Heywood (1947), and Backer (1951).

The assumptions basic to the hydraulic theories have been discussed by Carman (1941, 1948) and Klyachko (1948). According to them, the premises of that theory are:

(i) no pores sealed off;
(ii) pores distributed at random;
(iii) pores reasonably uniform in size;
(iv) porosity not too high;
(v) diffusion (slip) phenomena absent;
(vi) fluid motion occurring like motion through a batch of capillaries.

The simplest approach to the present problem is once more by construction of geometrical models. Leïbenzon (1947, pp. 31 ff.) gave a good review of such attempts. Thus Slichter (1899) considered a bed of spherical particles as equivalent to a porous medium. On this basis he set m in (6.3.1.1) equal to the diameter δ of the spheres and deduced the following expression for the permeability:

$$k = \frac{n^2 \delta^2}{96(1-P)}, \qquad (6.3.1.2)$$

where n is the same quantity as has been defined on page 18 (cf., Leïbenzon, 1947, p. 34). It is equal to 0.0931 for the densest packing of the spheres. Graton and Fraser (1935), Missbach (1938), and Dubinin (1941) extended the work on such models consisting of spheres. The modern version of the hydraulic radius theory, however, is based upon a completely different theoretical representation of porous media.

6.3.2. The Kozeny Theory. Kozeny's work (1927a) is today the most widely accepted explanation for the permeability as conditioned by the geometrical properties of a porous medium, although rather severe objections against it will be pointed out later. The Kozeny theory represents the porous medium by an assemblage of channels of various cross-sections, but of a definite length. The Navier-Stokes equations are solved simultaneously for all channels passing through a cross-section normal to the flow in the porous medium. Finally, the permeability is expressed in terms of the specific surface of the porous medium which is, according to remarks in section 6.3.1, a measure of a properly defined (reciprocal) hydraulic radius. However, certain aspects are neglected, notably by *assuming that* in a cross-section "normal" to the channel *there is no tangential component of the fluid velocity*. The Kozeny theory, therefore, neglects the influence of conical flow in the constrictions and expansions of flow channels, just as capillaric model theories do. The Kozeny equation was developed independently a few years after Kozeny in America by Fair and Hatch (1933).

In detail, the Kozeny theory proceeds as follows. Assume that a stream tube has a cross-section F and a pore cross-section f with

$$f = PF, \qquad (6.3.2.1)$$

where P is, as usual, the porosity. Then one has for the volume of liquid passing during the time dt:

$$Qdt = FdsP, \qquad (6.3.2.2)$$

where Q is the volume of fluid passing during unit time and ds the line element traversed in the time dt. Now, if v_p is the average pore velocity and q the filter velocity, one has:

$$v_p f = qF = Q. \qquad (6.3.2.3)$$

From (6.3.2.2) and (6.3.2.3) follows, if $v_p = ds/dt$,

$$q/v_p = f/F = P. \qquad (6.3.2.4)$$

As outlined earlier, the last equation is known as the Dupuit-Forchheimer equation. In the Kozeny theory it comes out as a consequence of the Kozeny-model of porous media and is, therefore, subject to the criticisms of that model as will be pointed out later. As an *a priori* assumption as postulated by Dupuit (1863), see section 6.2.1.

If one assumes a cross-section of the stream tube in the xy plane, the equation of motion is (cf. equation 2.2.2.1)

$$(\partial^2/\partial x^2 + \partial^2/\partial y^2) V_p = -\operatorname{grad} p/\mu, \qquad (6.3.2.5)$$

where V_p is the velocity in the point xy of the cross-section of the pores, and μ the viscosity. Introducing new variables

$$\xi = x/\sqrt{f} \quad \text{and} \quad \eta = y/\sqrt{f},$$

one obtains the differential equation

$$(\partial^2/\partial \xi^2 + \partial^2/\partial \eta^2) V_p = -f \operatorname{grad} p/\mu, \qquad (6.3.2.6)$$

which has the solution

$$V_p = f \frac{-\operatorname{grad} p}{\mu} \psi(x/\sqrt{f}, y/\sqrt{f}). \qquad (6.3.2.7)$$

The function ψ is given by:

$$\psi(x/\sqrt{f}, y/\sqrt{f}) = -(x^2 + y^2)/(4f) + \sum_{n=1}^{n=\infty} a_n \Phi_n(x + iy), \qquad (6.3.2.8)$$

where Φ denotes a harmonic function. Assuming that the fluid sticks to the walls of the pores, one has the following boundary condition:

$$\psi(x/\sqrt{f},\ y/\sqrt{f}) = 0, \qquad (6.3.2.9)$$

or,

$$(x^2 + y^2)/(4f) = \sum_1^\infty a_n \Phi_n(x + iy). \qquad (6.3.2.10)$$

If equation (6.3.2.10) is written in polar co-ordinates r and ϕ, one obtains for the boundary $(r = \rho)$:

$$\frac{\rho^2}{4f} = \sum_1^\infty a_n \Phi_n(\rho e^{i\phi}). \qquad (6.3.2.11)$$

If the circumference of the cross-section of the pore be M, and the radius of a circle of same circumference be ρ_m, with

$$\rho_m = \frac{M}{2\pi},$$

then it is possible to rewrite (6.3.2.8), observing the boundary condition, and one obtains:

$$\psi(r\cos\phi/\sqrt{f},\ r\sin\phi/\sqrt{f})$$
$$= \frac{M^2}{16\pi^2 f}\left(\sum_1^\infty a_n \Phi_n(re^{i\phi})\frac{4f}{\rho_m^2} - \frac{r^2}{\rho_m^2}\right) \equiv \frac{M^2}{16\pi^2 f}A. \qquad (6.3.2.12)$$

Then one obtains for the average pore velocity

$$v_p = \frac{\int V_p dF}{\int dF} = -\frac{f\ \mathrm{grad}\ p}{\mu F 16\pi^2}\frac{M^2 F}{f}A_m \qquad (6.3.2.13)$$

where

$$A_m = \iint A r\, dr\, d\phi / \iint r\, dr\, d\phi$$

is the mean value of A averaged over the area. If the expansion of the circumference is represented by $u_0 = M^2/f$, then it is clear that A and v_p become smaller, if u_0 becomes larger. In the limiting case of maximum expansion of the circumference, $v_p = 0$, since all points of the cross-section will be on its edge. Equally, the average value A_m becomes smaller, if the expansion of the circumference becomes larger; it must decrease faster than the expression $16\pi^2 c/u_0$, where c is a dimensionless number which depends only on the shape of the cross-section. The last condition is satisfied by the expression $16\pi^2 c/u_0^\zeta$ for all $\zeta > 1$.

Now, the pore cross-section of a virtual stream tube consists of a certain number n of single pore cross-sections, whose area f_1 we assume as equal to all others. Then

$$f = nf_1 \quad \text{and} \quad M = nM_1,$$

where M_1 is the circumference of one single pore.

If we write equation (6.3.2.13) in the following form

$$v_p = -\operatorname{grad} p F W / \mu, \qquad (6.3.2.14)$$

then this expression must not change if the number of the pores is changed, i.e., if we change n. According to the formulas above, one has

$$W = \frac{f}{F} c \frac{1}{u_0^{\zeta-1}} = \frac{c}{F} \frac{f^\zeta}{M^{2\zeta-2}} = \frac{c}{F} \frac{f_1^\zeta}{M_1^{2\zeta-2}} \frac{1}{n^{\zeta-2}}. \qquad (6.3.2.15)$$

It is obvious from this equation that W is invariant with respect to n, if the exponent ζ is made equal to 2.

Yet the filter velocity q, so far, is dependent on n. We have the equation

$$v_p = -\operatorname{grad} p \left(\frac{c}{\mu}\right)\left(\frac{f}{F}\right)\left(\frac{f}{M^2}\right) F, \qquad (6.3.2.16)$$

and also the equation

$$q = -\frac{k}{\mu}\operatorname{grad} p = \frac{f}{F} v_p = -\frac{c}{\mu}\operatorname{grad} p \left(\frac{f}{M}\right)^2 \left(\frac{f}{F}\right). \qquad (6.3.2.17)$$

Since f/M is independent of f/F and the latter constant along a stream line, the expression f/M must also be constant along a stream line and hence in a stream tube. Thus, in the case of a cylindrical stream tube of constant cross-section F, the quantities f and M must also be constant. However, the surface S_t in a stream tube of length L is given by

$$S_t = ML.$$

We can rewrite this as follows:

$$\frac{f}{M} = \left(\frac{f}{F}\right)\left(\frac{LF}{ML}\right) = \frac{P}{S},$$

where S is now the specific surface of the tube. Thus, we can write for (6.3.2.17):

$$q = -\frac{cP^3}{\mu S^2}\operatorname{grad} p. \qquad (6.3.2.18)$$

Comparing this with Darcy's law we obtain for the permeability

$$k = \frac{cP^3}{S^2}. \qquad (6.3.2.19)$$

This relation, called the "Kozeny equation," shows that the filter velocity is inversely proportional to the square of the surface area per unit volume. The number c fluctuates, theoretically, only very little. We have for a circle: $c = 0.50$; for a square: $c = 0.5619$; for an equilateral triangle: $c = 0.5974$; and for a strip: $c = \frac{2}{3}$. The number c is called "Kozeny constant."

One can now extend the Kozeny equation by introducing a "tortuosity" as an undetermined factor, in accordance with what has been done in the case of capillaric models. Thus, one can argue that equation (6.3.2.18) should refer to a "reduced" pressure gradient

$$\text{grad } p_{\text{red}} = \frac{1}{T} \text{ grad } p,$$

rather than to the "apparent" one, indicating that the actual flow path is T times longer than the "apparent" path straight across the porous medium; T, then, is called "tortuosity." This changes the expression for the permeability (6.3.2.19) to read:

$$k = \frac{cP^3}{TS^2}. \qquad (6.3.2.20)$$

It may be noted that, owing to a difference in the models employed, the tortuosity T enters into the expression for the permeability in the Kozeny theory in a different manner than in the capillaric model theory (see equation 6.2.4.4).

The Kozeny theory is intended to have general applicability to all porous media, because the constants c and T, theoretically, involve only the detailed structure of the medium. However, it should be noted that the concept of tortuosity is really alien to the Kozeny theory. It does not come into play during the actual deduction of the Kozeny equation and can therefore be justified only *a posteriori* by the desire to have another arbitrary parameter.

6.3.3. Modifications of the Kozeny Theory. The basic treatment of Kozeny has been modified on several occasions. In these modifications, the result that the permeability is essentially proportional to $1/S^2$, that is proportional to the square of a properly defined hydraulic radius, remains, but a great uncertainty exists as to the correct porosity factor. Equations analogous to (6.3.2.20) are commonly called "Kozeny" equations, whether or not the porosity dependence is that given by Kozeny. These various different porosity factors are usually obtained from more or less refined models.

Such modifications of the Kozeny equation have been proposed by Terzaghi (see Leïbenzon, 1947, p. 38), Zunker (1953), and Bakhmeteff and Feodoroff (1937). A much used modification was postulated by Carman (1937, 1938a, 1939a) who set (Kozeny-Carman equation)

$$k = \frac{P^3}{5S_0^2(1-P)^2} \qquad (6.3.3.1)$$

where S_0 is Carman's "specific" surface exposed to the fluid; i.e., the surface exposed to the fluid per unit volume of *solid* (not porous) material. Equation (6.3.3.1) implies that the value of the Kozeny constant c is equal to $\frac{1}{5}$ as this gives, according to Carman (1938), the best agreement with experiments. This is at variance with the calculations of Kozeny, according to which one ought to expect c to equal $\frac{1}{2}$. Here, therefore, is one of the facts which casts some doubts upon the Kozeny theory.

The latest modification of the hydraulic radius theory appears to be due to Sullivan (1942) who introduced an orientation factor θ, which is defined as the average value of the square of the size of the angle between a normal to the walls forming the microscopic flow channel and the macroscopic direction of flow. The resulting formula is

$$k = \frac{c\theta P^3}{S_0^2(1-P)^2}, \qquad (6.3.3.2)$$

where c is a Kozeny constant that should be the same for all channels of the same geometric shape and that should not vary markedly from shape to shape.

Further discussions of the hydraulic radius theory may be found in papers by Wiggin, Campbell and Maass (1939), Mitton (1945), Zhuravlevi and Sỹchev (1947), Baver (1949), Klyachko (1948), Arthur et al. (1950), Iberall (1950), and Loudon (1952). In these papers, the hydraulic radius theory has been carried to a high degree of refinement. Notably, Klyachko gives an analytical method of calculating the Kozeny constant. The theory has also been applied extensively to determine the internal structure of porous media, as will be shown later.

6.3.4. Experiments in Connection with the Hydraulic Radius Theory. Many experiments have been based upon the Kozeny theory. Mostly, they are determinations of the constants occurring therein, particularly of the internal surface of porous media. The surface determination will be discussed later in detail, and we shall confine ourselves in this section to the discussion of other experiments.

An actual substantiation of the Kozeny equation is almost impossible because it contains three parameters (S, c, T), all of which would have to be determined by independent means in order to obtain a valid check. This, however, introduces many errors, especially in view of the fact that T is not even properly defined as it was introduced by the Kozeny equation itself.

Nevertheless, a number of experiments have been performed which claim to "substantiate" the Kozeny theory, usually simply by measuring some of the constants occurring therein and showing that the values obtained are not unreasonable. Such experiments have been reported by Bartell and Osterhof (1928), Donat (1929), Kozeny (1932), Zunker (1932), Carman (1939a, 1939b), Wiggin, Campbell and Maass (1939), Fowler and Hertel (1940), Rigden (1943), Adams, Johnson and Piret (1949), Blaine and Valis (1949), Kwong et al. (1949), Rose and Bruce (1949), Coulson (1949, 1950), and Hoffing and Lockart (1951).

An improvement over the above "verifications" of the Kozeny equation has been attempted by Wyllie and co-workers (Wyllie, 1951; Wyllie and Spangler, 1952; Wyllie and Gregory, 1955). Here an attempt was made to attach a more precise physical meaning to the "tortuosity." Wyllie postulated that the electrical and fluid flow paths should be identical if the electrical and pressure potential were analogous. Thus, the tortuosity could be measured by independent means. However, inspection of Wyllie's work shows that there is some uncertainty regarding the proper expression of tortuosity in terms of the electrical analogue, reflecting itself in the fact that the Kozeny constant is, in every case, adjusted *a posteriori*.

The Kozeny theory, therefore, contains some vague factors and it is the opinion of the writer that more than a qualitative description of the phenomena cannot be expected from it.

6.4. Structure Determinations of Porous Media

6.4.1. Determination of Surface Area.
The Kozeny equation contains as a significant variable the specific surface area of a porous medium. Therefore it should permit one to calculate the specific surface area of a porous medium from permeability measurements, provided the other "Kozeny constants" are known. For purposes of surface determination, it is generally assumed that $c = \frac{1}{5}$ (in equation 6.3.2.19) and the tortuosity factor is absent. This corresponds to the "Kozeny-Carman equation" (6.3.3.1).

The determination of the surface area of porous media, mainly of industrial powders, has been chiefly advanced upon this basis by Carman (1938b, 1939b, 1939c) and by Lea and Nurse (1939). The method has since become very popular and many applications and refinements have been proposed. One of these refinements is to include a "slip" term in the Kozeny equation as will be discussed in Part VII of this study. Reviews of these methods have been given by Blaine (1941), Zimens (1944), and Svensson (1949).

Specific applications and apparatuses are discussed as follows: to sand and crushed particles, by Dalla Valle (1938), and by Arthur *et al.* (1950); to pulp and fibres, by Sullivan (1942), Sullivan and Hertel (1942), Pfeiffenberger (1946), Elting and Barnes (1948), Robertson and Mason (1949), and Brown (1950); to leather, by Mitton (1945); and finally to powders, by Keyes (1946), Pecover (1946), Rigden (1947), Carman and Malherbe (1950, 1951), Dodd, Davis and Pidgeon (1951), and Arakawa, Arakawa and Suito (1952). Some of the values obtained from these investigations are shown in table II.

Very few of the above papers contain any critical evaluations of the accuracy of the method. The Kozeny (-Carman) equation is usually assumed as unquestionable and the results are presented as final values obtained by means of that equation.

6.4.2. Determination of Other Geometrical Quantities.
According to the simple geometrical models discussed earlier in Part VI, permeability measurements can be used, theoretically at least, to determine other geometrical quantities of porous media. This idea had been taken up by Traxler and Baum (1936) who reported

a determination of "average pore size" from permeability measurements. Similarly, Gooden and Smith (1940) reported calculations of the average particle diameter of powders from permeability; Kuhn (1946) claims to have determined the gel structure in a similar manner; and Wilson (1953) uses the Kozeny equation to determine a mechanism of agglomeration.

These determinations are all based on rather specific assumptions about the porous medium, i.e., on some particular model. As the porous media under actual investigation were probably rather remote from such models, not too much confidence can be placed in the reported results.

6.5. Criticism of Hydraulic Radius Theory; Other Theories

6.5.1. Criticism of the Kozeny Theory.

Some criticisms of the Kozeny theory from a theoretical standpoint have already been mentioned during the exposition of it above. In addition, Coulson (1949) made an extensive investigation of the data of previous authors, and on the basis of information obtained, from fluid flow through systems of packed spheres, prisms, cubes, cylinders, and plates, he concluded that more general formulations still must be developed to describe adequately flow phenomena in non-spherical systems. Similarly, Childs and Collis-George (1950) put forth severe criticisms of the Kozeny theory from other theoretical reasoning. They state that it seems to be generally admitted that hydraulic-radius theories utterly fail to describe structured bodies such as, for example, "stiff-fissured" clays. The structural fissures contribute negligibly both to porosity and to specific surface, and yet they dominate the permeability. Again, neither the porosity nor the permeability are directed quantities, and therefore the Kozeny formula cannot indicate anisotropic permeability, which nevertheless seems to be the rule rather than the exception in nature.

Many discrepancies with the Kozeny theory can be obtained from experiments, provided not too many undetermined "fudge" factors are introduced into it. Thus, Kozeny himself (1927b), in an attempt to substantiate his formula experimentally, found discrepancies between calculated and measured surface areas of -69 to $+86$ per cent. It is a matter of taste whether one wants to consider these as merely experimental errors. More recently, Macey (1940) found enormous changes in permeability with porosity which are not explainable by the Kozeny equation. Similarly, Sullivan and Hertel (1940) and Sullivan (1941) showed that Kozeny's law breaks down, at least in the case of highly porous, fibrous media. Walas (1946) found that a modification in the Kozeny equation is necessary if it is to be applied to filtration, and Adamson (1950) observed that the Kozeny constant has to be adjusted *a posteriori* in every experiment, which is, of course, quite unsatisfactory. Wyllie and Rose (1950), in an attempt to justify the Kozeny equation, place much emphasis upon the "tortuosity," which, however, is a rather vague physical concept, and, if measured electrically, is in need of a proper justification for an application to fluid flow. Furthermore, Brooks and Purcell (1952), again claiming that their results substantiate the Kozeny equation, in fact find a large discrepancy when areas are measured by the gas-adsorption method and with the aid of the Kozeny equation. Finally, Kraus and co-workers (Kraus and Ross, 1953; Kraus, Ross and Girifalco, 1953) attempted to correlate pressure drops and flow rates through beds of many materials with gross particle surface, as determined from geometric considerations, and with microstructure surface from nitrogen adsorption measurements. They found that both surfaces contribute to pressure loss, but by *different* mechanisms. Childs and Collis-George (1950) summarize the consequences of these experiments appropriately by stating that although most authors claim agreement with observed values of permeability, "the fact seems to be that it has been impossible to secure a sufficiently wide range of variation of porosity and pore-size distribution to test the formulas severely, and the tests which have been reported above indicate a wide range of error. That the method of determining the specific area of powders by measurement of permeability has achieved considerable popularity (see sec. 6.4.1) may reflect only the essential similarity of most industrial problems involved."

It seems indicated, therefore, that investigations of theories of permeability must be based upon assumptions other than the hydraulic radius theory.

6.5.2. Drag Theories of Permeability. An approach different from that of Kozeny to physical explanation of permeability has been initiated by Emersleben (1924, 1925). This approach might be called "drag theory" of permeability. In it, the walls of the pores are treated as obstacles to an otherwise straight flow of the viscous fluid. The drag of the fluid on each portion of the walls is estimated from the Navier-Stokes equations, and the sum of all drags is thought to be equal to the resistance of the porous medium to flow (i.e., equal to μ/k, according to Darcy's law). It is to be expected that the drag theory will give good results in the case of highly porous media, such as fibres, where the single particles can be actually regarded as solitary within the fluid. A modern exposition of the drag theory has been given by Iberall (1950). Similar lines of thought have been followed recently by Brinkman (1947, 1948, 1949) and by Mott (1951).

Following Iberall (1950), the drag theory uses the model of a random distribution of circular cylindrical fibres of the same diameter and accounts for the permeability on the basis of the drag on individual elements. It is assumed that the flow resistivity of all random distributions of the fibres per unit volume will not differ, and that it is the same as that obtained with an equipartition of fibres in three perpendicular directions, one of which is along the direction of macroscopic flow. It is further assumed that the separation between fibres, and the length of individual fibres are both large compared to the fibre diameter (high porosities), and that the disturbance due to adjacent fibres on the flow around any particular fibre is negligible.

If it be assumed that fluid inertial forces are negligible (low local Reynolds number), an equation can be derived by equating the pressure at two planes perpendicular to the direction of macroscopic flow to the viscous drag force on all elements between the planes. It is assumed that the pressure drop necessary to overcome the viscous drag is linearly additive for the various fibres, whether parallel or perpendicular to the flow.

The drag force per unit length of a single fibre surrounded by similar fibres all oriented along the direction of flow and with moderate separations has been given approximately by Emersleben (1925):

$$f = 4\pi\mu v_p, \qquad (6.5.2.1)$$

where f is the drag force per unit length of fibre and v_p is the velocity of the fluid stream distant from the filament, i.e., the pore velocity.

If it is assumed that there are n filaments of unit length per unit volume, and that $n/3$ filaments are arrayed in each of the three perpendicular directions, the total drag force in a unit volume due to the $n/3$ filaments parallel to the flow can be equated to the pressure drop per unit length, so that

$$\Delta p/L = (4\pi n/3)\mu v_p. \qquad (6.5.2.2)$$

From calculations of the flow resistance for a cylinder perpendicular to a stream (see Lamb, 1932), the drag force for such filaments is given by

$$f = \frac{4\pi}{2 - \log \operatorname{nat} Re} \mu v_p \qquad (6.5.2.3)$$

in which log nat Re is the natural logarithm of the local Reynolds number, defined as $\delta v_p \rho/\mu$, with $\delta =$ fibre diameter and $\rho =$ fluid density.

The drag force on each of the two sets of $n/3$ filaments per unit volume, arrayed perpendicularly to the fluid flow, can be equated to the pressure drop per unit length, giving

$$\frac{\Delta p}{L} = \frac{4\pi n}{3(2 - \log \operatorname{nat} Re)} \mu v_p. \qquad (6.5.2.4)$$

Linear superposition or simple addition of the three sets of pressure drops necessary to overcome the drag of the three sets of filaments results in the total pressure gradient of

$$\frac{\Delta p}{L} = \frac{4\pi n}{3} \frac{4 - \log \operatorname{nat} Re}{2 - \log \operatorname{nat} Re} \mu v_p. \qquad (6.5.2.5)$$

The number of fibres n per unit volume, which is also equal to the total fibre length per unit volume, may be eliminated, as the apparent density of a fibrous pack in vacuum ρ_p is equal to the product of the fibre volume and the fibre density ρ_f or

$$\rho_p = \frac{\pi}{4}\delta^2 n \rho_f. \tag{6.5.2.6}$$

Eliminating n, equation (6.5.2.5) reduces to

$$\frac{\Delta p}{L} = \frac{16\mu v_p}{3}\frac{\rho_p}{\rho_f \delta^2}\frac{4 - \log\text{ nat } Re}{2 - \log\text{ nat } Re}. \tag{6.5.2.7}$$

The velocity profile between fibres is assumed to be sufficiently flat for high porosities so that the velocity may be taken as constant. The velocity v_p is therefore related to the volumetric rate of flow Q and the macroscopic cross-sectional area of a porous medium A by the relation (Dupuit-Forchheimer)

$$v_p = Q/(PA) = q/P, \tag{6.5.2.8}$$

$$q = Q/A. \tag{6.5.2.9}$$

It follows from the definition of the porosity P that

$$1 - P = \frac{\rho_p}{\rho_f}. \tag{6.5.2.10}$$

With the use of these two relations, equation (6.5.2.7) may be put in the form

$$\text{grad } p = q\,\frac{16\mu}{3}\frac{1-P}{P\delta^2}\frac{4 - \log\text{ nat }[\delta q \rho/(\mu P)]}{2 - \log\text{ nat }[\delta q \rho/(\mu P)]}. \tag{6.5.2.11}$$

Although the derivation, as given, assumed an incompressible fluid, it can be readily shown that the derived equations are unchanged for a compressible fluid, flowing isothermally, if the volumetric flow at the arithmetic mean pressure q_m is used in equation (6.5.2.11). It is therefore applicable to both liquids and gases.

If equation (6.5.2.11) is compared with Darcy's law, one obtains formally for the permeability:

$$k = \frac{3}{16}\frac{P\delta^2}{1-P}\frac{2 - \log\text{ nat }[\delta q \rho/(\mu P)]}{4 - \log\text{ nat }[\delta q \rho/(\mu P)]}. \tag{6.5.2.12}$$

This shows a very important point contiguous to the drag theory of permeability. It is noted that the permeability is not a constant but varies with flow velocity. This slow variation of permeability with flow is quite characteristic of many instances in which the flow is nominally viscous. While in general it is reasonable to assume that this effect is associated with fluid inertia, it is often difficult to account for it precisely. It is thus seen that the drag theory of permeability is not only a very different account of the latter, but even modifies the Darcy equation and thus leads to more general flow equations.

§6.5 CRITICISM OF HYDRAULIC RADIUS THEORY

As pointed out above, a different treatment of the drag theory of permeability has been proposed in a series of papers by Brinkman (1947, 1948, 1949). Brinkman assumes that the particles in the fluid are spheres of radius R and that they are kept in position by external forces, as in a bed of closely packed particles which support each other by contact.

Following Brinkman (1949), in order to formulate an equation describing the flow of a fluid through such a swarm, one has to consider the forces acting on a volume element of fluid containing many particles.

On the one hand there are normal and shearing stresses which act on the surface of the volume element. If no particles were present, these stresses would cause a force $F_1 dV$ which is given by the Navier-Stokes equation (2.2.2.1):

$$F_1 = -\operatorname{grad} p + \mu \operatorname{lap} v \qquad (6.5.2.13)$$

where p is, as usual, the pressure, v the flow velocity vector, and μ the viscosity. The inertia terms have been neglected (i.e., the density ρ of the fluid has been assumed as equal to zero), and the fluid has been assumed as incompressible.

On the other hand, owing to the presence of particles, equation (6.5.2.13) has to be modified. In the first place, there may be particles occupying positions near the surface of the volume element dV and diminishing the area of contact between the fluid inside and outside of the volume element. In the second place, if the particles are not closely packed, they may rotate and thus cause an increase in F_1. At any rate, the velocity of flow will be a rapidly fluctuating function of the position in the fluid, having zero value at the surface of the particles and maxima in the spaces between them.

In anticipation of a consistent application of statistical mechanics to the hydrodynamics in porous media (see section 6.5.3), Brinkman (1949) defined a mean flow velocity by taking an average over a volume V_0 which is small compared to the total volume V but which contains many particles:

$$q = \frac{1}{V_0} \int_{V_0} v \, dV. \qquad (6.5.2.14)$$

The particles which are kept in position by external forces exert a damping force on the velocity of flow of the fluid. It seems reasonable to assume that this force $F_2 dV$ is proportional to the mean velocity and to the viscosity of the fluid. Brinkman therefore put

$$F_2 = -(\mu/k)q \qquad (6.5.2.15)$$

where k is a constant depending on the particle density and their radii to be discussed below.

Consequently, if only viscous forces are taken into account, F_1 and F_2 are the only forces acting on the volume element considered. For a stationary state of flow one therefore has

$$F_1 + F_2 = 0. \qquad (6.5.2.16)$$

As an approximation Brinkman substituted the expressions given in equations (6.5.2.13) and (6.5.2.14) into equation (6.5.2.16). Expression (6.5.2.13) will only be a good approximation for a low particle density on account of the disturbing influence of the particles as discussed before. Thus, for low particle densities Brinkman put

$$-\operatorname{grad} p + \mu \operatorname{lap} q - \frac{\mu}{k} q = 0. \qquad (6.5.2.17)$$

This equation may of course be extended in the usual way to include inertial terms. It is seen that the present theory, again, leads to a modification of Darcy's law through the inclusion of the term $\mu \operatorname{lap} q$. For high particle densities the term $\mu \operatorname{lap} q$ is negligible compared to $\mu q/k$. This means that Darcy's law is the limiting form of equation (6.5.2.17) for low permeabilities.

Equation (6.5.2.17), therefore, is a new description of the state of flow for both low and high particle densities. The permeability k in equation (6.5.2.17) has yet to be determined. Following Brinkman, an expression for this permeability k may be found by calculation of the damping force F_2 as defined by equation (6.5.2.15) for an assemblage of spheres of radius R. It may be assumed that this force is the resulting sum of the viscous forces exerted by the individual spheres on the fluid. It is found first by calculating the force exerted by the fluid on one of the spheres and then by summing up over all the spheres.

The fluid flow round one sphere is derived from equation (6.5.2.17). This means that the influence of the other particles is taken into account by the damping term in equation (6.5.2.17). Brinkman emphasized that this treatment might be justified if the surrounding spheres were small compared to the central sphere, and may be applied by analogy to the case where all the spheres are of the same size. Thus, for an incompressible fluid a solution of equation (6.5.2.17) has to be found which obeys the condition of incompressibility

$$\operatorname{div} q = 0, \qquad (6.5.2.18)$$

and certain boundary conditions at the surface of the particle. These boundary conditions are that the normal and the tangential velocity components at the surface of the particle should be zero. The condition for the normal component results from equation (6.5.2.18) in the usual way. The condition for the tangential component results from certain considerations on the continuity of the viscous stresses combined with an assumption on variation of the permeability k near a boundary. Brinkman emphasized that the boundary condition for the tangential component can only be derived from some additional assumptions on the swarm of particles.

The actual condition chosen by Brinkman is consistent with the simplest assumptions. Calculation of the detailed flow pattern through the swarm of spheres might lead to a justification of equation (6.5.2.17) combined with the above boundary conditions. Brinkman stated, however, that the result might well be that the mean rate of flow near a boundary could not be represented by the theory discussed above.

With these boundary conditions a solution of equation (6.5.2.17) and (6.5.2.18) can be found which represents a uniform parallel state of flow infinitely far removed from the sphere (Brinkman, 1947). By calculating the viscous forces, the force F on the sphere is eventually found to be equal to:

$$F = 6\pi\mu q_0 R \left[1 + \frac{R}{\sqrt{k}} + \frac{R^2}{3k}\right], \qquad (6.5.2.19)$$

where q_0 is the mean rate of flow far from the sphere and R is its radius.

Equation (6.5.2.19) is a modification of Stokes' law as generalized for a dense swarm of spheres. For an isolated sphere (i.e., the permeability approaching infinity), the correction factor approaches the value 1.

An expression for the permeability can now be found by calculating the total damping force on a unit volume of the swarm containing n spheres.

This force is equal to nF, where F is given by equation (6.5.2.19). But, on the other hand, by definition it is equal to $\mu q_0/k$. From this equality Brinkman finally found for k:

$$k = \frac{R^2}{18}\left[3 + \frac{4}{1-P} - 3\sqrt{\frac{8}{1-P} - 3}\right], \qquad (6.5.2.20)$$

$$\frac{4}{3}\pi n R^3 = 1 - P, \qquad (6.5.2.21)$$

where P is, as usual, the porosity.

6.5.3. Statistical Theories. In spite of the partial success with the physical explanation of Darcy's law, the theories presented heretofore must be termed unsatisfactory upon general grounds. All these theories, essentially, make models of the porous medium which are ordered and regular to such an extent that the Navier-Stokes equations can be solved for the latter. However, in general, natural porous media are extremely disordered so that it seems a rather poor procedure to represent them by something which is intrinsically *ordered*. It has therefore been suggested that a model of a porous medium should be made which is intrinsically *disordered* and thus the problem would be approached from an opposite extreme. It is of course quite hopeless to try to solve the Navier-Stokes equations for such disordered media; methods of statistical mechanics have to be applied instead.

Hubbert (1940) was probably the first to apply some statistics to the theory of flow through porous media, in that he showed that the motion parallel to grad ϕ must be proportional to the magnitude of that same gradient. This he deduced by statistical considerations, *viz.*, by proving that forces normal to grad ϕ

are dropping out, and by integrating the others. These considerations are based upon the concept that the microscopic flow velocity of each fluid particle is proportional to the acting force.

Another very elementary application of statistics of flow through porous media has been undertaken by Childs and Collis-George (1950), who base their theory on pore size distribution. The latter is treated by a simple statistical theory, based upon the calculation of the probability of occurrence of pairs of pores of all the possible sizes, and of the contribution of each such pair to permeability.

Finally, statistics have been applied more consistently to flow through porous media by Taub (1951) and by Scheidegger (1954). Taub (1951) developed equations on the basis of the Maxwell-Boltzmann distribution function of gases. Density, pressure, and velocity are defined in accordance with kinetic theory, and a one-particle system equivalent to a many-particle system is considered. The particle is treated as undergoing a Markov process. The macroscopic quantities, density, velocity, pressure satisfy the classical partial differential equation describing the flow of fluids in permeable media.

Similarly, Scheidegger (1954) applied the statistics of disordered phenomena as exemplified by Einstein's theory of the Brownian motion to the flow of fluids through porous media. The flow of each "point" of the fluid is treated as a statistical process which, however, is not defined as a stochastic process but as a statistical process within an ensemble of macroscopically (not microscopically) identical porous media. The main outcome of this procedure is that it leads to a modification of Darcy's law, in that it introduces a new macroscopic quantity which is termed "dispersivity." This quantity is indicative of the sideways dispersion which a stream of fluid is known to undergo. The differential equations of flow are now formulated in terms of a probability distribution function for each "point" of the fluid which has to be properly defined.

The method of statistics can be demonstrated most easily if the following assumptions are made:

(a) The porous medium is isotropic and, averaged over the ensemble, homogeneous.

(b) The external forces on the fluid are homogeneous and time independent—this may be realized by a constant pressure drop along a given direction.

(c) If (a) and (b) are assumed, then the time averages of the displacements per unit time taking place in one sample of the ensemble, are equal to the displacements per unit time averaged over the ensemble. This is often called "ergodic hypothesis" and actually tied up with the hypothesis of disorder.

If we consider a "particle" of fluid, i.e., a volume of fluid which is small enough not to be separated into separate channels during its journey through the porous medium, we will be able to talk about the position of this particle at a given time.

§6.5   Criticism of Hydraulic Radius Theory

We split the time from 0 to t into N small intervals τ such that

$$N\tau = t. \tag{6.5.3.1}$$

In every time interval τ the particle will undergo a displacement $\boldsymbol{\xi}$ (with components ξ, η, ζ with respect to a Cartesian co-ordinate system). The probability that a given displacement $\boldsymbol{\xi}$ will take place in a time interval τ, shall be denoted by $v(\boldsymbol{\xi})$. By this definition, $v(\boldsymbol{\xi})$ is a time-averaged probability. Because of the ergodic hypothesis, this probability is assumed to be equal to the ensemble-averaged probability. Under the assumptions made, this probability taken over the ensemble will depend neither on the position of the particle in the porous medium nor on the instant. The probability $v(\boldsymbol{\xi})$ is assumed to be normalized so that

$$\int v(\xi, \eta, \zeta) d\xi d\eta d\zeta = 1. \tag{6.5.3.2}$$

The average (over the ensemble) displacement in the time interval τ will be denoted by $\bar{\boldsymbol{\xi}}$; it is given by

$$\bar{\boldsymbol{\xi}} = \int \boldsymbol{\xi} v(\boldsymbol{\xi}) d^3\xi = (\bar{\xi}, \bar{\eta}, \bar{\zeta}). \tag{6.5.3.3}$$

It is very awkward to deal with probability distributions whose medians are not zero, so that it will be convenient to introduce a co-ordinate system in which the median of the probability distribution will be zero. We shall indicate this co-ordinate system by primes. Thus we set:

$$\boldsymbol{\xi}' = \boldsymbol{\xi} - \bar{\boldsymbol{\xi}}. \tag{6.5.3.4}$$

In this primed system, the probability for the fluid particle to have a displacement $\boldsymbol{\xi}'$ will be:

$$w(\boldsymbol{\xi}').$$

The average of $\boldsymbol{\xi}'$ is zero. The functions $w(\boldsymbol{\xi}')$ may depend on $\bar{\boldsymbol{\xi}}$ as a parameter.

The fundamental question is now: What is the probability $w_N(x', y', z')$ over the ensemble that a fluid particle with initial co-ordinate $(0, 0, 0)$ in the primed system will have co-ordinates x', y', z' after the time $t = N\tau$.

Because of the assumptions made, viz., that the porous medium is homogeneous and isotropic, and the field of forces homogeneous, the original probability density $w(\xi', \eta', \zeta')$ will be split into a product of three identical probabilities:

$$w(\xi', \eta', \zeta') = w(\xi')w(\eta')w(\zeta') \tag{6.5.3.5}$$

and, furthermore, these probabilities will be identical for every time step τ.

Thus, in terms of probability calculus, the problem is nothing but the composition of repeated trials. No matter what the original distribution w is, the central limit theorem states that the distribution after N trials is Gaussian,

provided N is large. Thus, if we denote by σ the standard deviation of w, it follows:
$$w_N(\xi', \eta', \zeta') = (2\pi N\sigma^2)^{-\frac{3}{2}} \exp\{-(x'^2 + y'^2 + z'^2)/(2N\sigma^2)\}. \quad (6.5.3.6)$$

The quantities σ and τ are constants during the motion (although σ may depend on $\vec{\xi}$) such that one may set
$$\sigma^2/\tau = 2D \quad (6.5.3.7)$$
where D is some "factor of dispersion." Using $N\tau = t$, yields:
$$w(t, x', y', z') = (4\pi Dt)^{-\frac{3}{2}} \exp\{-(x'^2 + y'^2 + z'^2)/(4Dt)\} \quad (6.5.3.8)$$
and, returning to un-primed co-ordinates:
$$v(t, x, y, z) = (4\pi Dt)^{-\frac{3}{2}} \exp\{-[(x-\bar{x})^2 + (y-\bar{y})^2 + (z-\bar{z})^2]/(4Dt)\}. \quad (6.5.3.8)$$

This is the fundamental probability distribution describing the journey of a fluid parcel through the porous medium. This particular relation has been obtained under rather stringent assumptions. In particular, the hypothesis that there is no correlation between the ith and the $i+1$st step is certainly an oversimplification. However, it would not be difficult to take this correlation into account, as is being done in the statistical theory of turbulence.

The next task is to find a connection between the average displacement $\bar{x}$ and the external field of forces. If the latter are assumed as homogeneous and time-independent, then the laws of viscous flow state that the displacement $\vec{\xi}$ is for every small time-step τ:
$$\bar{x}(t = N\tau) = N\vec{\xi}; \quad (6.5.3.9)$$
$$|\vec{\xi}|/\tau = (B/\mu)|\text{grad } p|\cos\theta. \quad (6.5.3.10)$$

Here, grad p denotes the external force and $\cos\theta$ the angle between the vectors $\vec{\xi}$ and grad p; B is a certain factor indicative of the reciprocal resistance of the opening through which the fluid (of viscosity μ) moves during the time step τ.

Equation (6.5.3.10) will now be averaged over the ensemble. The average $\vec{\xi}$ will be in the direction of $-$ grad p if the porous medium is assumed to be isotropic. Its magnitude will be equal to the average of the component of $\vec{\xi}$ in the direction of $-$ grad p; i.e., equal to the average of $|\vec{\xi}|\cos\theta$. Thus one has
$$\vec{\xi}/\tau = -(\bar{B}/\mu)\,\text{grad }p\,\overline{\cos^2\theta}$$
or
$$\boldsymbol{v}_p = -(\bar{B}/\mu)\,\text{grad }p\,\overline{\cos^2\theta} \quad (6.5.3.11)$$
where $\boldsymbol{v}_p$ is now the pore-velocity vector. Therefore, the probability distribution $v(\boldsymbol{x}, t)$ can now be expressed as follows:
$$v(\boldsymbol{x}, t) = (4\pi Dt)^{-\frac{3}{2}} \exp-\left\{\frac{[x + t(\bar{B}/\mu)\,\text{grad }p\,\overline{\cos^2\theta}]^2}{4Dt}\right\}. \quad (6.5.3.12)$$

The average position of the median $\tilde{x}$ of this distribution is:

$$x(t) = -\bar{B}\,\mathrm{grad}\,p\,t\,\overline{\cos^2\theta}/\mu \tag{6.5.3.13}$$

and the average of the square-length $\overline{x^2}$ is (from the general theory of Gauss-distributions):

$$\overline{x^2(t)} = 6Dt + \left(\frac{\bar{B}}{\mu}\,\mathrm{grad}\,p\,\overline{\cos^2\theta}\,t\right)^2. \tag{6.5.3.14}$$

One can evaluate the factor D in terms of other quantities. In order to do this, one has two possibilities: a dynamic procedure, and a geometric procedure.

(a) *Dynamic procedure.* Newton's law of motion for a small particle of the fluid is

$$B(f - \rho\ddot{x}) = \mu\dot{x} \tag{6.5.3.15}$$

where f is the force per unit volume (apart from the viscous forces). If (6.5.3.15) is multiplied by x,

$$Bfx - B\rho\ddot{x}x = x\dot{x}\mu. \tag{6.5.3.16}$$

Since

$$\ddot{x}x = (d/dt)(x\dot{x}) - \dot{x}^2 = (\tfrac{1}{2})(d^2/dt^2)(x^2) - \dot{x}^2, \tag{6.5.3.17}$$

this yields:

$$Bfx - \frac{1}{2}B\rho\frac{d^2}{dt^2}(x^2) + B\rho\dot{x}^2 = \frac{\mu}{2}\frac{d}{dt}(x^2). \tag{6.5.3.18}$$

This equation can be averaged over the ensemble; the average of fx is simply $\bar{f}\tilde{x}$; the average of f, then, is:

$$\bar{f} = -\,\mathrm{grad}\,p\,\overline{\cos^2\theta} \tag{6.5.3.19}$$

since only the ordered component of f remains through the averaging process. Thus, averaging (6.5.3.18) yields:

$$\left(\frac{\bar{B}}{\mu}\,\mathrm{grad}\,p\,\overline{\cos^2\theta}\right)^2 t - \frac{1}{2}\frac{\bar{B}\rho}{\mu}\frac{d^2}{dt^2}\overline{x^2(t)}$$

$$+ \frac{B\rho}{\mu}\left(\frac{\bar{B}}{\mu}\,\mathrm{grad}\,p\,\overline{\cos\theta}\right)^2 = \frac{1}{2}\frac{d}{dt}\overline{x^2(t)}. \tag{6.5.3.20}$$

This is a differential equation for $\overline{x^2(t)}$ whose stationary solution is

$$\overline{x^2(t)} = 2\frac{\bar{B}\rho}{\mu}\left[\left(\frac{\bar{B}}{\mu}\,\mathrm{grad}\,p\,\overline{\cos\theta}\right)^2 - \left(\frac{\bar{B}}{\mu}\,\mathrm{grad}\,p\,\overline{\cos^2\theta}\right)^2\right]t$$

$$+ \left(\frac{\bar{B}}{\mu}\,\mathrm{grad}\,p\,\overline{\cos^2\theta}\right)^2 t^2 \tag{6.5.3.21}$$

Comparing this with (6.5.3.14), one finally obtains:

$$D = \frac{\bar{B}\rho}{3\mu}\left[\overline{\left(\frac{B}{\mu}\operatorname{grad} p \cos\theta\right)^2} - \left(\frac{\bar{B}}{\mu}\operatorname{grad} p \overline{\cos^2\theta}\right)^2\right]. \quad (6.5.3.22)$$

In this last equation, $\bar{B}$ and θ are given by the porous medium. Therefore, the expression for D may be written as follows

$$D = \frac{\rho a'}{\mu^3}(\operatorname{grad} p)^2 \quad (6.5.3.23a)$$

where now a' contains everything that depends on the porous medium. It is, therefore, a constant of the porous medium and may be termed the latter's *dynamical dispersivity*.

In a similar fashion, we can contract everything which in equation (6.5.3.13) refers to the porous medium only, into one single constant k/P. We thus obtain:

$$q/P = \boldsymbol{v}_p \equiv \bar{\boldsymbol{x}}(t)/t = -(k/\mu P)\operatorname{grad} p. \quad (6.5.3.24)$$

This corresponds to Darcy's law (which is thus shown to be valid for the median of the probability distribution); it is therefore also in the present theory appropriate to call k "permeability" of the porous medium.

The dynamic method is applicable if there is enough time in each flow channel for complete mixing by molecular sideways diffusion to take place. This is the condition which is necessary so that each small cross-section of a flow channel can be assumed as a unit, so as to enable one to write down Newton's law of motion in the form of (6.5.3.15).

(b) *Geometric procedure.* An alternative method to calculate the factor of dispersion is obtained by the remark that, if there is no interchange of particles pertaining to adjacent streamlines (incidentally, this corresponds to *true* laminar flow), then the geometrical distribution of v must be independent of $\bar{\boldsymbol{v}}_p$. This means that v as a function of $\boldsymbol{x}$ and $\bar{\boldsymbol{x}}$ must be independent of the mean flow velocity (although, in a transient state, this distribution will be reached at an earlier time if the velocity is increased). The condition for this is that the exponent in (6.5.3.12) is independent of $\bar{\boldsymbol{v}}_p$ if t is eliminated by means of (6.5.3.24). One obtains:

$$D = a''\frac{k}{\mu P}|p|\operatorname{grad} \quad (6.5.3.23b)$$

where a'' is another constant of the porous medium—preferably called its *geometrical dispersivity*. Equation (6.5.3.23b) is applicable if there is no appreciable molecular sideways diffusion from one streamline into another.

The above statistical theory refers to homogeneous forces only. It is, however, easy to effect a generalization to non-homogeneous forces by observing that the fundamental probability distribution (6.5.3.12) is a solution of the differential equation:

$$\frac{\partial v}{\partial t} = \text{lap}\,(Dv) + \text{div}\left[\bar{B}v\,\text{grad}\,p\,\overline{\frac{\cos^2\theta}{\mu}}\right]. \qquad (6.5.3.25)$$

Inserting the above-found values for $\bar{B}$, etc., we find:

$$\frac{\partial v}{\partial t} = \text{lap}\,(Dv) + \text{div}\left[v\frac{k}{\mu P}\,\text{grad}\,p\right], \qquad (6.5.3.26)$$

which is also valid for non-homogeneous forces.

Finally, one has to formulate the continuity condition. At time $t = t_0$ there will be a certain distribution of the fluid in the porous medium. Let this distribution be denoted by $w(x_0, t_0)$ where the position co-ordinate is now denoted by x_0. Naturally, x_0 has the same range of values as x. At time t (arbitrary), the "original" distribution of fluid $w(x_0, t_0)$ will have the probability-density $v(x, x_0; t, t_0)$ to be at the spot x at time t. Here, $v(x, x_0; t, t_0)$ is a "key" solution of the differential equation (6.5.3.26). It satisfies the boundary condition and is a delta-function for $t = t_0$ of the argument $x - x_0$. It has, therefore, properties similar to a Green's function. Thus, the distribution of fluid at time t will be given by $w(x, t)$ as follows:

$$w(x, t) = \int v(x, x_0; t, t_0) w(x_0, t_0) d^3 x_0. \qquad (6.5.3.27)$$

The physical concept of continuity states that the distributions w must be proportional to the densities of the fluid times the local porosity and thus equation (6.5.3.27) can be written as follows:

$$P(x)\rho(x, t) = \int v(x, x_0; t, t_0)\rho(x_0, t_0)P(x_0)d^3x_0 \qquad (6.5.3.28)$$

This is the required continuity condition.

The continuity condition in the form of (6.5.3.28) is rather awkward to deal with. It would be preferable to have it in differential form rather than in integral form. This aim can be achieved by taking the time-derivative with respect to t on both sides of equation (6.5.3.28):

$$P(x)\frac{\partial \rho(x, t)}{\partial t} = \int \frac{\partial v(x, x_0; t, t_0)}{\partial t}\rho(x_0, t_0)P(x_0)d^3x_0. \qquad (6.5.3.29)$$

However, the solution v is subject to the differential equation (6.5.3.26). One can thus insert $\partial v/\partial t$ from (6.5.3.26) into (6.5.3.29) to obtain:

$$P(x)\frac{\partial \rho(x, t)}{\partial t} = \int \left\{\text{lap}\,(Dv) + \text{div}\left[v\frac{k}{\mu P}\,\text{grad}\,p\right]\right\}\rho(x_0, t_0)P(x_0)d^3x_0 \qquad (6.5.3.30)$$

Upon letting $t = t_0$, this yields:

$$P \frac{\partial \rho}{\partial t} = \text{lap} \, [P\rho D] + \text{div} \left[\rho \frac{k}{\mu} \text{grad} \, p \right]. \qquad (6.5.3.31)$$

This is the macroscopic equation which describes "dispersivity." If the factor D is assumed equal to zero, the fluid motion is identical to that described by Darcy's law. However, if this factor is not set equal to zero, then a macroscopic effect, dispersion, occurs. Individual particles of the fluid not only move along streamlines resulting from Darcy's law, but they are also dispersed sideways.

In the particular statistical theory which is given in detail in this section, the effect of dispersion of particles is expressed by quantities denoted by a', a'' and called "dispersivity." This "dispersivity" appears in the equations of motion and turns out to be a constant of the porous medium, but only because certain rather restrictive assumptions had been made. The above theory must therefore be held correct only in a qualitative sense. It could be considerably refined, for example, by assuming correlations between the various time steps. Nevertheless, statistics of disorder seem to be the only sensible approach to the phenomenon of flow through porous media.

Verifications of the above theory have been obtained in connection with experiments on miscible displacement in porous media. They will be discussed in section 8.7.3.

From the above expositions it is seen that theories of single phase flow through porous media, which are different from hydraulic radius theories, in general also introduce a modification of the Darcy equation. However, in all cases they seem to describe effects which are known to exist and which are not properly accounted for by the Darcy equation. More discussion will be devoted to such effects in Part VII.

6.5.4. Analogy of Laminar Flow in Porous Media with Turbulent Flow in Bulk Fluids. The dispersion of the individual particles during laminar flow in porous media has also been interpreted in a different manner (Yuhara 1954). The fluctuations of the velocities of the particles can be claimed as being analogous to the fluctuations of velocity during eddy motion in turbulent flow. Thus, the flow path of a particle in flow through a porous medium can be regarded as analogous to the trajectory of a particle in turbulent hydraulic motion. It might be expected, therefore, that methods yielding reasonable results in the case of turbulent flow through pipes, etc., will also yield reasonable results if applied to the laminar flow through porous media.

Indeed, the flow path of a fluid particle in an ensemble of macroscopically identical porous media is a probability-phenomenon (over the ensemble) and therefore could be thought of as corresponding to the random flow path of a fluid particle during turbulent motion in a bulk mass of fluid. The appearance of a "dispersivity" in the hydrodynamics in porous media may be thought of as

§6.5 CRITICISM OF HYDRAULIC RADIUS THEORY

analogous to the diffusivity of eddies, etc., in turbulent motion. It is by this analogy with turbulence that diffusivity equations have been suggested for the laminar flow through porous media without recourse to a proper statistical treatment.

It should be noted, however, that the analogy between laminar flow in porous media and turbulent motion in bulk masses of fluid is not very complete. This has not always been realized in applying it; moreover, it has even been claimed that the statistical hydrodynamics in porous media would suffer from a major flaw because it does not use the same velocity correlation tensors as are used in the statistical theory of turbulence.

Thus, let us introduce velocity correlation tensors in the statistical hydrodynamics of laminar flow in porous media and investigate where this leads.

In analogy with the statistical theory of turbulence, we shall therefore consider the local (pore) velocity v as fundamental dynamical variable. The commonly considered "correlation tensor" is then defined as follows, the summation convention being applied (Scheidegger, 1956):

$$R_{ik}(r) = \overline{[v_i(x) - \bar{v}_i(x)][v_k(x+r) - \bar{v}_k(x+r)]} \tag{6.5.4.1}$$

Assuming narrow channels, the law of viscous flow (corresponding to the Hagen-Poiseuille equation) can be written as follows:

$$v_i = -\frac{b_{ik}}{\mu}\frac{\partial p}{\partial x_k}, \tag{6.5.4.2}$$

expressing the fact that, in narrow channels, the flow is proportional to the driving force. Assuming a homogeneous pressure gradient, the mean velocity can therefore be expressed as follows

$$\bar{v}_i = -\frac{1}{\mu}\frac{\partial p}{\partial x_k}\bar{b}_{ik} = -\frac{1}{\mu}k_{ik}\frac{\partial p}{\partial x_k} \tag{6.5.4.3}$$

where

$$k_{ik} = \bar{b}_{ik} \tag{6.5.4.4}$$

turns out to be proportional to the permeability tensor (see sec. 4.3.3). Thus, the deviation of the velocity from its mean is

$$v_i - \bar{v}_i = -\frac{1}{\mu}\frac{\partial p}{\partial x_k}(b_{ik} - \bar{b}_{ik}). \tag{6.5.4.5}$$

Finally, forming the correlation tensor, we get

$$R_{ik}(r) = \frac{1}{\mu^2}\frac{\partial p}{\partial x_l}\frac{\partial p}{\partial x_m}$$
$$\times \overline{\{b_{il}(x)b_{km}(x+r) - b_{il}(x)\bar{b}_{km}(x+r) - b_{km}(x+r)\bar{b}_{il}(x) + \bar{b}_{il}(x)\bar{b}_{km}(x+r)\}}. \tag{6.5.4.6}$$

If we assume that the porous medium is isotropic, this formula can be simplified to yield

$$R_{ik}(\boldsymbol{r}) = \frac{1}{\mu^2} \frac{\partial p}{\partial x_l} \frac{\partial p}{\partial x_m} b_{iklm} \qquad (6.5.4.7)$$

where, for abbreviation, one has set

$$b_{iklm} = \overline{b_{il}(\boldsymbol{x})b_{km}(\boldsymbol{x}+\boldsymbol{r})} - k_{il}k_{km}. \qquad (6.5.4.8)$$

As is easily seen, the tensor b_{iklm} is a tensor which depends on the porous medium only. It is obtained by forming the required averages. Equations (6.5.4.7–8) show that the velocity correlation tensor in laminar flow through porous media depends only in a trivial way on the dynamical variables (*viz.*, on the pressure). The main dependence of this tensor is on the porous medium. It thus turns out that the velocity correlation tensor is unsuitable as a dynamic variable, and that the analogy between turbulent flow and flow through porous media is not very complete, in spite of the fact that the contrary has been claimed. It should be noted, however, that this result is mainly due to the particular statistical model of porous media which was under discussion here. It is therefore still possible that correlation tensors might yield something useful in a more complicated theory.

PART VII

GENERAL FLOW EQUATIONS

7.1. Limitations of Darcy's Law

7.1.1. High Flow Rates.
In the previous parts of this review, Darcy's law has been accepted as fundamental for the deductions reached. However, it has been mentioned above that Darcy's law possibly can be valid only in a certain "seepage" velocity domain, outside which more general flow equations must be used to describe the flow correctly.

In order to characterize this seepage velocity domain, it is customary to introduce a "Reynolds number" Re as follows:

$$Re = q\rho\delta/\mu \tag{7.1.1.1}$$

where q is the (scalar) filter velocity, ρ the density, μ the viscosity of the fluid, and δ a diameter associated with the porous medium, i.e., the average grain or pore diameter or some other length corresponding to the hydraulic radius theory. In this and the following formulas it is always understood that consistent units (i.e., c.g.s. units) are used.

Similarly, it has been customary to express the resistance to flow of the porous medium by the introduction of a "friction factor" λ as follows:

$$\lambda = 2\delta \operatorname{grad} p/(q^2\rho) \tag{7.1.1.2}$$

where, in addition to the previous symbols, $\operatorname{grad} p$ is the one-dimensional pressure gradient (a suitable modification could be made to account for gravity flow if so desired). In this representation, Darcy's law would be written as follows:

$$\lambda = C/Re \tag{7.1.1.3}$$

where C is a constant. It is therefore seen that the flow is subject to Darcy's law if λ and Re (for one porous medium) are inversely proportional.

The representation of the flow by means of Reynolds number and friction factor, of course, is dependent on the choice of the length δ. It is thus tied up with the notion of a hydraulic length as introduced by the hydraulic radius theory. We have listed above objections to the hydraulic radius theory; one of them was that the diameter δ is not really defined, either geometrically or by statistical reasoning. This same objection holds against the present representation. However, all experiments on the limitations of Darcy's law (at least at the high velocity end) invariably seem to be tied up with the notion of Reynolds number. Results obtained in this manner, therefore, must be regarded as subject to the same principal shortcomings as the hydraulic radius theory, including the Kozeny equation.

Nevertheless, many investigations have been directed towards finding the range of Reynolds numbers for which Darcy's law is valid. That the representation by Reynolds numbers was chosen originally, is of course owing to the assumption of an analogy between the flow in tubes and the flow in a porous medium: i.e., the latter is thought to be equivalent to an assemblage of capillaries. Therefore one has been looking for a phenomenon in porous media similar to the onset of turbulence in tubes, which takes place at a definite Reynolds number. It was expected that above a certain number, which would be universal for all porous media, deviations from Darcy's law would occur.

Experiments for the determination of this "critical" Reynolds number have been reviewed, for example by Romita (1951), Hudson and Roberts (1952), and Leva and co-workers (1951). Original investigations have been made by a great number of people, notably Zunker (1920), Ehrenberger (1928), Fancher and Lewis (1933), Fancher, Lewis and Barnes (1933), Hickox (1934), Schaffernak and Dachler (1934), Bakhmeteff and Feodoroff (1937, 1938), Carman (1937), Hatch (1938), Brieghel-Müller (1940), Chardabellas (1940), Gustafson (1940), Nissan (1942), Grimley (1945), Ruth (1946), Khanin (1948), Fox (1949), Cambefort (1951), Nielsen (1951), and Plain and Morrison (1954). The account of these researches in section 7.2.1 is essentially based upon the articles by Romita 1951) and by Leva et al. (1951).

In these investigations, a great discrepancy regarding the "universal" Reynolds number, above which Darcy's law would no longer be valid, is evident. The values range between 0.1 (Nielsen, 1951) and 75 (Plain and Morrison, 1954). The uncertainty of a factor 750 about the critical Reynolds number may reflect in part the actual indeterminacy of the diameter δ; however, it is probably essentially due to a fundamental failure of the hydraulic radius theory of porous media. Some attempts have also been made to obtain a more definite value for the critical Reynolds number by somehow bringing the porosity (which is a dimensionless quantity) into the expression for the former (cf. Bakhmeteff and Feodoroff, 1938; Brownell and Katz, 1947) but without much success, as it appears. A very interesting result of these experiments is, however, that the critical Reynolds number above which turbulence is believed to occur is much lower for porous media than for straight tubes (where it is around 2000; cf. sec. 2). Again, the conclusion from this can only be that a porous medium is *not* equivalent to an assemblage of straight tubes.

The net result of these experiments is that above a "certain" value for the seepage velocity, Darcy's law is no longer valid. A universal characterization of this "certain" value, however, has not been achieved. Attempts to accomplish this by introducing a Reynolds number which would become "critical" above a certain value, universal for all porous media, have failed because this critical value was found to vary erratically within a factor 750 for various porous media and fluids. Even at that, the critical Reynolds number would be at least about 25 times smaller than would be expected if a porous medium were equivalent to an assemblage of straight "tubes."

7.1.2. Molecular Effects. Deviations from Darcy's law have not only been observed at high flow rates, for which they would be expected by analogy with flow in tubes. A category of such deviations is particularly manifest in the flow of gases. Here, the deviations are presumably due to molecular effects.

Thus, Fancher, Lewis and Barnes (1933, 1934) and Manegold (1937) observed that air permeabilities are higher than liquid permeabilities in the same porous medium as calculated from Darcy's law. In fact, this is an indication that Darcy's law is not valid for gases. Similar findings were obtained by Brown and Bolt (1942) and by Ruth (1946). A theory of this phenomenon is given later (cf. sec. 7.3.1). As far as a characterization of the point when Darcy's law becomes no longer valid for gas flow is concerned, Calhoun and Yuster (1946) summarized the facts by stating that Darcy's law breaks down if the pore diameters become comparable with, or less than, the molecular mean free paths of the flowing gas. One can again observe an analogy between this breakdown of Darcy's law with the breakdown of Hagen-Poiseuille's law in capillaries: If the radius of the capillary is made smaller and smaller, the originally viscous flow of the gas undergoes a transition to slip flow and thence to molecular streaming as observed by Knudsen (cf. sec. 2). Of course, such an analogy is possible only within the light of the hydraulic radius theory of permeability and is therefore subject to the corresponding limitations.

Deviations from Darcy's law that might be explained by an analogy with Knudsen's findings for capillaries are not the only ones, however. There are other deviations that might also be due to molecular effects.

Comparing air and liquid permeabilities for a series of porous media, Calhoun (1946) found that neither air nor liquid permeabilities were constant as calculated from Darcy's law. Air and different liquid permeabilities were not in agreement: the average for liquids was usually somewhat lower than for air. Both liquid and air permeabilities depended either upon mean pressure or upon pressure gradient. Similarly, Grunberg and Nissan (1943) claim to have found that there is no correlation between the permeability of porous media to gases and to liquids. Permeabilities to gases depend mainly on the mean linear speed of the gas, whereas permeabilities to liquids also depend on the pore diameter and the specific surface tension. A similar study has been made by Deryagin and Krylov (1944). On the other hand, Bulkley (1931) was unable to detect deviations from Darcy's law for various fluids for pore diameters as small as $5.6\,\mu$.

In making experiments with one only fluid, Idashkin (1936) found that the gas permeability increases with gas pressure. This is an effect opposite to that expected from analogy with Knudsenian flow in capillaries. Strange "anomalies" were also observed by Bodman (1937) when experimenting with the flow of water through saturated soils. One particular result was that the permeability may change with time. Similarly, Grisel (1936) observed a change of permeability with time when passing water through cement.

Explanations that might account for the observed effects will be discussed systematically in section 7.3.2. Apart from deviations which have already been accounted for by the drag theory of permeability (Brinkman, 1947; Iberall, 1950), it will be seen that adsorption, capillary condensation, and molecular diffusion can actually produce such "anomalies" (in the light of Darcy's law) as were listed above. In connection with such theories, some further experiments will be mentioned which were performed to substantiate the explanations. The above list of observed anomalies is therefore not complete.

7.1.3. Other Effects. The limitations of Darcy's law due to turbulence and due to molecular effects, as outlined above, are not the only ones. A series of other possible effects can cause Darcy's law to break down.

One group of anomalies may be ascribed to what may be termed "boundary effect." This effect has been observed by Zhavoronkov et al. (1949), Aerov and Umnik (1950), Arthur et al. (1950), Morales et al. (1951), Schwartz (1952), and Schwartz and Smith (1953). The boundary effect causes the velocity of a fluid stream passing through a porous bed to decrease near the wall and the centre of the bed; it is at a maximum somewhere in between. This is particularly evident in tubes packed with pellets where the fluid velocity is seen to increase with distance from the centre of the tube until a maximum is reached about one pellet diameter from the wall and then to fall sharply. Explanations of this profile have been attempted based upon Prandtl's mixing length and upon the change of voidage with radial position.

Another group of anomalies with respect to Darcy's law can be ascribed to the presence of ions in the percolating fluid. Such anomalies have been observed in soils by Nayar and Shukla (1943a, 1943b, 1943c, 1949), Shukla and Nayar (1943), Shukla (1944), and Sillén and co-workers (Sillén, 1946, 1950a, 1950b; Sillén and Ekedahl, 1946; Ekedahl and Sillén, 1947). In connection with the analysis of rocks this effect has been observed by Urbain (1941), Breston and Johnson (1945), Griffiths (1946), Miller (1946), Miller et al. (1946), and Heid et al. (1950). A particularly fine set of experiments has been reported by von Engelhardt and Tunn (1954). The change of permeability with pH of the percolating solution is not a simple relationship: there are maxima and minima. It is generally thought that the permeability changes are brought about by an actual electrochemical reaction between the solution and the porous medium; in some instances (e.g., in the case of the rocks) clay swelling may also play a role.

There are indications, however, that ionic anomalies in the permeability of a porous medium are at least partly due to effects which are not at all chemical. Calhoun (1946) and Calhoun and Yuster (1946) analysed the ion exchange which is supposed to take place when salt solution is passed through porous media containing sodium ions, and found that there must be another effect besides mere ion exchange. Yuster (1946) made similar experiments with artifical quartz filters that do not contain any sodium or other ions at all, and was still able to observe an ionic effect. Ruth (1946), Bridgwater (1950), and Duriez (1952) gave an explanation of the phenomenon based upon the analogy with a similar effect observable in capillaries. When an ionic fluid is mechanically forced through a capillary, a small electromotive force, known as the streaming potential, can be detected. Under the influence of this self-generated potential, a small electro-osmotically transported flow of liquid will take place in the same manner as it would if the electromotive force were externally applied. This effects a retardation in the total flow. (Cf. Ruth, 1946.) The corresponding ionic effect in porous media is therefore often termed "electro-osmosis." Whether electro-osmosis is able to account for the total permeability anomalies observed with ionic fluids is uncertain. In general, it probably occurs concurrently with the above-mentioned chemical changes.

7.2. Equations for High Flow Velocities

7.2.1. Heuristic Correlations.

The high velocity flow phenomena occurring in porous media, described in section 7.1.1, can be put into mathematical terms in several ways. Without attempting to understand the physics of the effect, one can simply try to fit heuristically curves or equations to the experimental data so as to obtain a correlation between pressure drop and flow velocity.

In this instance, Forchheimer (1901) suggested that Darcy's law be modified for high velocities by including a second-order term in the velocity:

$$\Delta p/\Delta x = aq + bq^2 \qquad (7.2.1.1)$$

and later by adding a third-order term:

$$\Delta p/\Delta x = aq + bq^2 + cq^3. \qquad (7.2.1.2)$$

Here, the pressure drop correlation is written for the linear case (linear dimension x); furthermore, p is as usual the pressure (gravity is neglected), q is the seepage velocity, and a, b, c are thought to be constants. The Forchheimer equations were postulated from semi-theoretical reasoning by analogy with the phenomena occurring in tubes. The third-order term was added to make the equation fit experimental data better. Nevertheless, the Forchheimer relations are valid only if the seepage velocity is very high.

Another heuristic correlation has been postulated by White (1935), who set

$$\Delta p/\Delta x = aq^{1.8}; \qquad (7.2.1.3)$$

this result has been derived from an analysis of dry air flow through packed towers. The value of the exponent, however, seems not too well established, and thus Missbach (1937) set

$$\Delta p/\Delta x = aq^m, \qquad (7.2.1.4)$$

with m indetermined between 1 and 2. He further stated that as the diameter of the beds used for the packings became smaller, m approached 1 and the equation became Darcy's law. Similar experiments which bear out that there is a non-linear correlation between pressure drop and percolation velocity have also been made by Wodnyanszky (1938), Spaugh (1948), Linn (1950), and Uguet (1951). The exact value of the exponent m, however, seems to vary from case to case so that no universal correlation could be achieved.

It is therefore obvious that a different type of correlation has to be established if there is to be any hope for it to be universal. An extremely popular correlation is that between Reynolds number Re and friction factor λ (cf. equations 7.1.1.1 and 7.1.1.2 for their definition). Unfortunately, all such correlations are subject to limitations as the Reynolds number significantly depends upon a definition of a pore diameter which cannot be achieved properly. Furthermore, if a true physical significance is to be attached later to correlations based upon the Reynolds number, one has to introduce all the assumptions of the hydraulic radius theory.

A good review of the correlations that have been attempted between friction factor and Reynolds number has been given by Romita (1951). Other general discussions of the subject have been given by Kuz'minȳkh and Apakhov (1940), Rose (1945a), Verschoor (1950a), Leva (1949), Leva et al. (1951), and Hudson and Roberts (1952).

The number of papers proposing correlations between friction factor and Reynolds number is very great. Fancher and Lewis (1933) represented the results of experiments as plots between λ and Re; their data were, however, still principally in the viscous range. Uchida and Fujita (1934); (see also Fujita and Uchida, 1934) passed gases through beds of broken limestone, lead-shot, and Raschig rings and expressed their results in the form of the equation:

$$\operatorname{grad} p = \rho A \left(\frac{q^2}{2\delta}\right)^r Re^s \left(\frac{\Delta}{\delta}\right)^t \qquad (7.2.1.5)$$

where r and t are functions of the packing, and A and s are functions of both packing and Reynolds number; Δ is the diameter of the vessel holding the packed bed. Experimental limits of $r, s, \ldots$ were also supplied.

Further investigations were carried out by Lindquist (1933). By a careful investigation of previous results, this author came to the conclusion that Darcy's law is valid for $Re < 4$; for $4 < Re < 180$ he postulated the equation

$$\lambda Re = aRe + b \qquad (7.2.1.6)$$

with $a = 40$ and $b = 2500$. Givan (1934) made similar investigations and came up with an analogous relationship, but with the values $a = 34.2$ and $b = 2410$.

Similar investigations were undertaken by Fair and Hatch (1933), Mach (1935, 1939), Meyer and Work (1937), Johnson and Taliaferro (1938), Iwanami

(1940), and Kling (1940). Kling obtained the result that Darcy's law holds for up to $Re = 10$; for $10 < Re < 300$ he postulated the correlation:

$$\lambda = 94/(Re)^{0.16}.$$

Similar correlations have been postulated by Veronese (1941), namely:

$$\left. \begin{array}{c} \tfrac{1}{4} Re \lambda = 1150 \text{ for } Re < 5 \\ \tfrac{1}{4} Re\, 0.73 \lambda = 720 \text{ for } 5 < Re < 200 \\ \tfrac{1}{4} \lambda = 15.5 \text{ for } Re > 200 \end{array} \right\} \qquad (7.2.1.7)$$

The last expression is not valid for $Re > 2000$.

A series of further investigations has been made by Hancock (1942), Holler (1943), and Gamson et al. (1943). The last authors give plots of the λ-Re correlation for air flow over wet and dry spheres and cylinders. They obtained separate curves for wet and dry packings in tubes. Furthermore, Allen (1944) obtained relationships for the flow of air, naphtha, and mineral oil through beds of granular adsorbents. Finally, the subject of possible correlations has also been discussed by Schoenborn and Dougherty (1944), Taylor and Davies (1947), Morse (1949), Cambefort (1951), and Green and Duwez (1951).

The above-listed investigations are concerned with a correlation between λ and Re only. However, it should be expected that other variables also have an influence. The porosity should be especially significant, for, if the idea of a Reynolds number is maintained, it should obviously be the pore velocity and not the filter velocity whose value is significant for the onset of turbulence. According to the usual assumption (originally due to Dupuit), the pore velocity v_p is equal to q/P. Chilton and Colburn (1931a, 1931c) disagreed with this, however, and postulated that the porosity must be accounted for differently. Subsequently, Chalmers et al. (1932) introduced the porosity concept into the friction factor which they defined as follows:

$$\lambda_c = \text{grad } p \; \delta P^2/(\rho q^2). \qquad (7.2.1.8)$$

This expression for λ was used to obtain correlations.

A similar modification of the friction factor was introduced by Barth and Esser (1933) who wrote their friction factor λ_B as follows:

$$\lambda_B = \lambda P^3/(1 - P). \qquad (7.2.1.9)$$

With this friction factor, they postulated the following λ-Re correlation:

$$\lambda_B = 490/Re + 100/(Re)^{\frac{1}{2}} + 5.85. \qquad (7.2.1.10)$$

They claim that this correlation is valid for $5 < Re < 5000$.

Bakhmeteff and Feodoroff (1937, 1938) defined λ by the earlier convention (7.1.1.2), but correlated the values of λ they obtained for gas flow through beds of lead-shot by bringing the porosity explicitly into their equations. Thus for laminar flow ($Re < 5$) they wrote (see Leva et al., 1951):

$$\lambda = \frac{710}{ReP^{\frac{4}{3}}} \qquad (7.2.1.11)$$

and for turbulent flow ($Re > 5$)

$$\lambda = \frac{24.2}{Re^{0.2}P^{\frac{4}{3}}}. \qquad (7.2.1.12)$$

For flow of gases and liquids through porous carbon, Hatfield (1939) claimed that the logarithm of λP^2 (where λ is defined as in 7.1.1.2) could be linearly correlated with the logarithm of the Reynolds number for the flow range $10^{-5} < Re < 100$.

Further investigations to take the porosity effect into account were also made by Hatch (1940, 1943) and by Rose and Rizk (1948). Furthermore, Happel (1949) correlated $\lambda' = \lambda/(1-P)^3$ with the Reynolds number and reported that for the turbulent range

$$\lambda' = \text{const.}/(Re)^{0.22}. \qquad (7.2.1.13)$$

Finally, Romita (1951) also came up with a correlation taking the porosity effect into account, namely:

$$\lambda_r = 190/Re_r + 20/Re_r^{\frac{1}{2}} + 2.7 \qquad (7.2.1.14)$$

with

$$\lambda_r = \tfrac{2}{3}\lambda P^{\frac{7}{3}}/(1-P); \quad Re_r = \tfrac{1}{4}ReP^{\frac{1}{3}}/(1-P). \qquad (7.2.1.15)$$

Romita claims that these relationships would be valid for $4 < Re < 5000$.

There have also been attempts at correlations between the flow variables involving porosity and particle diameter δ which do not employ the representation by friction factor and Reynolds number. Burke and Plummer (1928) suggested the following flow equation

$$\text{grad } p = K[(1-P)/P^3]\rho q^2/\delta^2; \qquad (7.2.1.16)$$

and Zabezhinskiĭ (1939) postulated

$$\text{grad } p = Kq/\delta^n \qquad (7.2.1.17)$$

where $n = 1$ for high fluid velocities and $n = 2$ for low ones. In both equations, K signifies some constant. Finally, Hiles and Mott (1945), investigating the flow of gases through beds of coke and other granular materials, arrived at the following experimental correlation:

$$\text{grad } p = Kq^n \qquad (7.2.1.18)$$

where n and K are experimental constants which depend on the size of the granules. The value of these constants has been determined for a series of conditions.

§7.2 EQUATIONS FOR HIGH FLOW VELOCITIES

Just as porosity has been handled by various investigators in various ways, so the shape of the particles and other variables has been handled variously as factors influencing fluid flow (Leva et al., 1951). Wadell (1934) defined a shape factor for single particles as the ratio of the surface of a sphere having the same volume as the particle to the surface of the particle. Zeisberg (1919) had published pressure-drop data for various types of packing. Chilton (1938, p. 2211) converted these data, as well as the data of White (1935), to values of friction factors for the various shapes for use in his previously published (Chilton and Colburn, 1931c) equation. Blake (1922) correlated data on glass cylinders, Raschig rings, and crushed pumice by a linear plot on log-log co-ordinates of (grad $pP^3/[\rho q S]$) versus ($\rho q/[\mu S]$) where S is the value of surface area of packing per unit volume of packed tube (see Leva et al., 1951).

Similarly, Burke and Plummer (1928; see also Burke and Parry, 1935) studying the flow through spherical lead-shot in various diameter tubes, concluded that pressure drop is a function of a modified Reynolds number. They obtained

$$\text{grad } p = (K\rho q^2 S/P^3)(\mu S/\rho q)^{2-n} \qquad (7.2.1.19)$$

where n is a function of the Reynolds number and K again a constant.

Furnas (1931) reported on the effect of a large number of variables. However, he expressed his data in the form

$$\text{grad } p = \frac{A}{(\rho q)^B} \qquad (7.2.1.20)$$

where A and B are complex functions of particle size, bed porosity, and even of the fluid properties such as temperature, viscosity, density, and molecular weight. This, of course, is not very satisfactory.

Meyer and Work (1937) related the bed voidage for a given packing to some value P_n representing the loosest packing possible for the specified material and thus brought the notion of particle shape into their correlation.

Carman (1937) correlated the pressure drop data of other authors by the following dimensionally homogeneous formula (Leva et al., 1951):

$$\text{grad } p \frac{P^3}{\rho q^2 S_1} = C\left(\frac{\mu S}{\rho q}\right)^{0.1} \qquad (7.2.1.21)$$

where $S_1 = S + 4/\Delta$ (Δ is the diameter of the tube containing the packing and S the specific surface area) and C is a constant depending on particle shape. Oman and Watson (1944) correlated their pressure drop data in the "turbulent" flow range of air flow through (dense and loose-packed) beds of a variety of substances by refining the correlation of Burke and Plummer (1928). They adjusted the exponent of porosity occurring in the friction factor, so as to obtain the best correlation.

Rose (1945a) analysed literature data for non-spherical particles and claimed that they correlate with those for spherical particles by use of appropriate shape factors. Brownell and Katz (1947) also correlated pressure drop data of other investigators as well as their own data on air flow through salt beds. They arrived at the following equation (see Leva et al., 1951):

$$\text{grad } p = fq^2\rho/(2\delta P^n). \tag{7.2.1.22}$$

The factor f may be obtained from curves of Moody (1944) for flow through empty pipes as function of a Reynolds number which is defined as

$$Re_M = \delta\rho q/\mu P^m. \tag{7.2.1.23}$$

The exponents m and n are dependent on particle shape and bed porosity and are presented as experimentally derived curves. These investigations were later revised (Brownell et al., 1950) as it was found that the exponents m and n rose rapidly as the porosity approached unity. Therefore, in the revised correlation, the numerical values of the porosity functions are correlated directly, without assuming that the latter can be written as a simple power of P. The revised correlations use plots of these porosity functions against porosity with parameters of particle sphericity. Cornell and Katz (1951) used such relationships to predict the flow rates in industrial applications.

Finally, Leva and co-workers (Leva, 1947; Leva and Grummer, 1947; Weintraub and Leva, 1948; Leva et al., 1951) also investigated the possibility of correlations. In the last paper mentioned, they arrived at the following correlation

$$\text{grad } p = \text{const. } f\rho q^2\theta(1-P)/(\delta P^3) \tag{7.2.1.24}$$

where θ is a particle shape factor and f is called "modified friction factor." For smooth particles such as glass and porcelain, it is given by the relation

$$f = 1.75(\delta\rho q/\mu)^{-0.1}. \tag{7.2.1.25}$$

With rougher materials such as alundum or clay, it has the form

$$f = 2.625(\delta\rho q/\mu)^{-0.1}. \tag{7.2.1.26}$$

For still rougher packings, for example, aloxite or magnesium oxide granules, one finally has

$$f = 4.00(\delta\rho q/\mu)^{-0.1}. \tag{7.2.1.27}$$

Reviewing the correlations exhibited in the present section, it must be said that there is an extreme variety of them. It is quite certain that they cannot *all* be universally valid, as practically any one of them contradicts all others. We shall see later that this variety is probably due to the fact that there is actually no physical basis for an expectation that flows should be analogous if the Reynolds numbers are. The correlations, therefore, are at best valid each for an application to a set of very specialized porous media. In this instance, the correlations may be very useful for engineering applications to particular systems, but proper caution should be used if they are to be applied in any other setup than for which they had originally been obtained.

7.2.2. Theoretical Equations. The heuristic correlations of the flow variables in the "turbulent" flow region do not really shed any light upon the physics of the phenomenon. Not only are they based upon such vague concepts as the Reynolds number, but, moreover, the great variety of correlations that have been claimed to be valid also shows that the heuristic approach may at best be able to serve as a basis for calculation of flow rates in substances similar to those for which the correlations have been obtained. They seem to be quite useless, however, as a basis for further examination of the physical principles of the phenomenon.

One will therefore be prompted to attack the problem from a theoretical standpoint. For such an attack, the same methods are available as have been used for an investigation of the flow through porous media in the "laminar" region, i.e., where Darcy's law is valid. It is thus possible to apply dimensional analysis, or to extend either the hydraulic radius theory (i.e., the theory of capillaric models as well as the Kozeny theory) or the drag or statistical theories of permeability to turbulent flow.

A. *Dimensional analysis.* Thus, starting with dimensional analysis, we may note that, in fact, the representation of friction factors *versus* Reynolds number, discussed above, is an outcome of dimensional considerations. For the friction factor and the Reynolds number are both dimensionless groups and, therefore, one must be a numerical function of the other. Derivations of the form of this numerical function have been attempted only experimentally as was discussed in section 7.2.1. Now, it must be expected that there are more variables influencing the flow through porous media than were discussed heretofore. In this instance, Rose and co-workers (Rose, 1945a, 1945b, 1945c, 1949, 1951; Rose and Rizk, 1949) made a thorough study of the possible variables that might influence flow and the dimensionless combinations in which they might occur in a flow equation.

They assumed that the linear (filter) velocity q through a bed of granular material must depend somehow upon the density ρ and the viscosity μ of the fluid. It also must depend on the height h of the bed; the diameter δ of the particles constituting the bed; the diameter Δ of the container into which the bed is packed; the porosity P of the bed; and the gravity g. It may also be reasonably believed that the resistance to flow, and hence the difference of hydraulic head H across the bed, depends upon the height e of the surface roughness of the pores. In addition, the resistance to flow must vary with the shape of the particles of which the bed is composed and with their size distribution. However, a suitable definition will render these two variables expressible by two dimensionless quantities Z and U respectively. Thus, the relationship governing turbulent flow in porous media may be written symbolically as follows:

$$H = \Phi(q^\alpha h^\beta \delta^\gamma \rho^\zeta \Delta^\xi \mu^\theta g^\tau e^\eta P^\lambda Z^\sigma U^\omega) \tag{7.2.2.1}$$

where Φ denotes "a function of." If finally the dimensions of mass, length, and time of the variables are inserted into equation (7.2.2.1) and the exponents are equated, in compliance with the standard procedure of dimensional analysis, one obtains the following equation of flow for an incompressible fluid:

$$\frac{H}{\delta} = \Phi \left\{ \left(\frac{q\delta\rho}{\mu}\right)^\zeta \left(\frac{h}{\delta}\right)^\beta \left(\frac{\delta g}{q^2}\right)^\tau \left(\frac{\Delta}{\delta}\right)^\xi \left(\frac{e}{\delta}\right)^\eta P^\lambda Z^\sigma U^\omega \right\}. \tag{7.2.2.2}$$

Unfortunately, dimensional analysis is not able to yield more information than is expressed by equation (7.2.2.2). The values of the exponents α, β ... and the form of the function have to be determined experimentally. Rose and co-workers have made many experiments with this intent, but these run along the same lines as those discussed in section 7.2.1. No additional theoretical insight into the physics of the phenomenon can be gained in this manner.

With some additional assumptions, Leĭbenzon (1945a, 1945b) and Sokolovskiĭ (1949a, 1949b) deduced theoretically the form of one of the unknown quantities. The latter author wrote the flow equation in the following dimensionless form:

$$\text{grad } H = - \Phi(q)\mathbf{q}/q \tag{7.2.2.3}$$

where Φ is again "a function of" which has to be somehow determined.

Taking the steady state flow of incompressible fluids, one can formulate the continuity equation and it is seen that the latter is satisfied if

$$\Phi(q) = \frac{q}{K} \frac{1}{\sqrt{1 - \frac{q^2}{m^2}}}. \tag{7.2.2.3}$$

It is easy to see that for $m = \infty$, (7.2.2.3) is equivalent to Darcy's law. If m is finite, one obtains a deviation from Darcy's law and Sokolovskiĭ claims that the equation is good in the range $0 \leq q \leq m$. The constants m and K, of course, have still to be determined experimentally and may depend on other variables. Khristianovich (1940) developed an analytical method to find solutions of equations of the type of (7.2.2.3).

Finally Lapuk and Evdokimova (1951), in the case of two-dimensional seepage into a sink, established the following law by dimensional analysis:

$$q = CP^{(3n-\frac{1}{2})}\mu^{(1-2n)}\rho^{(n-1)}(dp/dr); \tag{7.2.2.4}$$

this relates the velocity of filtration q to the porosity P, absolute viscosity μ, density ρ, pressure p and radial distance from the sink r. Furthermore, C is a dimensionless coefficient related to the Reynolds number and n is called the "exponent" of the filtration law, and may also be dependent on the Reynolds number.

B. *Capillaric models*. The next possibility of deriving flow laws and of understanding them physically lies in the employment of capillaric models, i.e., the employment of analogies with flow through tubes. Such models have been mentioned, for example by Blake (1922), Lindquist (1933), Ward (1939), Brieghel-Müller (1940), Meinzer and Wenzel (1940), Radushkevich (1941), Ishikawa (1942), Takagi and Ishikawa (1942), and Wentworth (1946). Scheidegger (1953) gave a systematic discussion of such capillaric models. The law for turbulent flow in a tube is

$$dp = c' \frac{\rho v^2}{\delta} dx \qquad (7.2.2.5)$$

where v is the pore velocity and c' a certain constant. All the other symbols have the meaning defined in section 6.2 of this monograph. Turbulent flow will occur when the Reynolds number Re

$$Re = \delta v \rho / \mu \qquad (7.2.2.6)$$

exceeds a certain value, usually assumed to be in the neighbourhood of 2000. For smaller Reynolds numbers, the flow will be laminar as expressed by the Hagen-Poiseuille equation:

$$dp = -c \frac{\mu v}{\delta^2} dx. \qquad (7.2.2.7)$$

The application of the turbulent flow laws to porous media can now be sought in exactly the same fashion as the Hagen-Poiseuille equation was applied to models of porous media in section 6.2. Thus, starting with the first simple model discussed there, namely that of parallel capillaries, it is at once apparent that the flow in all capillaries will be either turbulent or laminar, according to the external pressure applied. For the turbulent region, a calculation can be made similar to that of section 6.2.2 for laminar flow. We have for the porosity, if there are n capillaries per unit area in each spatial direction,

$$P = \tfrac{3}{4} n \pi \bar{\delta}^2 \qquad (7.2.2.8)$$

and for the macroscopic velocity,

$$q = v \frac{\pi}{4} \bar{\delta}^2 n. \qquad (7.2.2.9)$$

It should be noted that, because of the particular form of this model, the Dupuit-Forchheimer assumption of $v = q/p$ is not valid. Thus, we obtain from equation (7.2.2.5):

$$\frac{dp}{dx} = 9 \rho c' \frac{q^2}{\delta P^2}. \qquad (7.2.2.10)$$

This model is obviously inadequate as no transition zone between laminar and turbulent flow can occur in it. Attention will have to be confined to models featuring different pore sizes.

Using first the parallel type model discussion in section 6.2.3, we obtain at once from equation (6.2.3.3)

$$q = -(\text{grad } p)^{\frac{1}{2}} (9c'\rho)^{-\frac{1}{2}} P \int_0^{\delta_R} \delta^{\frac{3}{2}} \alpha(\delta) d\delta$$

$$- \text{grad } p(P/(3\mu c)) \int_{\delta_R}^{\infty} \delta^2 \alpha(\delta) d\delta. \qquad (7.2.2.11)$$

Here, δ_R is defined as that value of δ which corresponds to the critical Reynolds number.

It is obvious that the last formula (7.2.2.11) is far remote from anything experimentally observed, for example the Forchheimer equation (7.2.1.1).

An attempt to introduce turbulence into the serial models discussed in section 6.2.4 is therefore indicated. The two laws of motion (7.2.2.5) and (7.2.2.7) entail

$$p_2 - p_1 = -\int_{\substack{\text{laminar}\\\text{region}}} \frac{c\mu v}{\delta^2} ds + \int_{\substack{\text{turbulent}\\\text{region}}} \frac{c'\rho v^2}{\delta} ds. \qquad (7.2.2.12)$$

inserting (6.2.4.8), (6.2.4.9), (6.2.4.10) yields (neglecting capillary pressure differentials):

$$\frac{p_2 - p_1}{x} = -q \frac{16c}{3\pi^2} \frac{\mu P}{n^2} \int_{\delta_R}^{\infty} \frac{\alpha(\delta) d\delta}{\delta^6} + q^2 \frac{64c'}{3\pi^3} \frac{\rho P}{n^3} \int_0^{\delta_R} \frac{\alpha(\delta) d\delta}{\delta^7} \qquad (7.2.2.13)$$

Again, n can be replaced by the tortuosity T by virtue of equation (6.2.4.15). We then obtain

$$\text{grad } p = -q \frac{3c\mu T^2}{P} \left\{ \int_{\delta_R}^{\infty} \frac{\alpha(\delta) d\delta}{\delta^6} \right\} \left(\int \delta^2 \alpha(\delta) d\delta \right)^2$$

$$+ q^2 \frac{9c'\rho T^3}{P^2} \left\{ \int_0^{\delta_R} \frac{\alpha(\delta) d\delta}{\delta^7} \right\} \left(\int \delta^2 \alpha(\delta) d\delta \right)^3 \qquad (7.2.2.14)$$

This expression appears to be almost of the form

$$\text{grad } p = Aq + Bq^2, \qquad (7.2.2.15)$$

which would correspond to the Forchheimer equation (7.2.1.1). However, one should keep in mind that δ_R is a function of q, according to (7.2.2.6). Only if it is assumed that the integrals do not depend much on the actual value of δ_R, equation (7.2.2.14) will "explain" the empirical formula of Forchheimer. Furthermore, the critical Reynolds number for the pore velocity has been taken as equal to about 2000. This is the Reynolds number at which turbulence occurs in a *straight* tube, and according to the present models, this should also be the Reynolds number at which turbulence would set in in the porous medium. In order to make a proper comparison, the pore velocity v should be expressed in terms of seepage velocity q. Using the Dupuit-Forchheimer assumption, one would have $q = vP$. However, it should be remembered that the Dupuit-Forchheimer assumption is not valid in the present models. If the flow channels are, as postulated, independent of each other in the three spatial directions, only $\frac{1}{3}$ of the porosity is available for flow in any one direction, and one has correspondingly $q = vP/3$. Thus, expressed in terms of seepage velocity, the critical Reynolds number will be $P/3$ times the original one. Thus, for a medium with porosity $P = 0.2$, turbulence should set in at a Reynolds number (calculated with q) equal to about 130. This is about fifty times too high if it is compared with the actual limits of Darcy's law. The only conclusion possible is that there is something seriously wrong with the models. The chief discrepancy comes, of course, from the fact that the critical Reynolds number is equal to 2000 only for *straight* tubes; the flow channels in an actual porous medium are far from straight. We have pointed out in Part II of this monograph that the critical Reynolds number is very much affected by any curvature of the tube, and therefore it must be expected that there is no such thing as a "universal" Reynolds number for porous media at which turbulence would set in. Depending on the curvature of the channels, the critical Reynolds number will be different in some flow channels than in others, even if their cross-sections are identical and even if the latter are put together to form porous media of identical porosity and "tortuosity." But with this, any basis of calculating δ_R (necessary in the above formulas) is logically non-existent. Moreover, any representation of porous media using a Reynolds number seems somewhat unsatisfactory. This verdict, of course, also applies to the heuristic correlations discussed in section 7.2.1. The discussion of capillaric models plainly bears out this fact, *viz.*, that there is no proper physical basis for assuming that the flows (and therefore friction factors) should be identical for identical Reynolds numbers, unless the curvature of the flow channels is somehow brought into the picture.

C. *Kozeny theory.* If the occurrence of turbulent flow in capillaric models is extended to the complexity achieved in the Kozeny-Carman theory, one arrives at a corresponding theoretical description of non-laminar flow in porous media. This idea has been chiefly followed up by Ergun and Orning (1949; see also Ergun, 1952a, 1953). These authors started from an equation postulated by Reynolds (1900) for the flow in linear tubes where the resistance offered by friction to the motion of the fluid is represented as the sum of two terms, proportional respectively to the first power of the fluid velocity and to the product of the density ρ of the fluid with the second power of its velocity, *viz.*:

$$|\text{grad } p| = av + b\rho v^2. \qquad (7.2.2.16)$$

The factors a and b are functions of the system. Ergun and Orning assume now that equation (7.2.2.16) is also valid for flow through porous media, with the filter velocity q instead of v, which again implies that the medium is assumed as equivalent to an assembly of capillaries. The authors then proceed to interpret the two terms as representing viscous and kinetic energy losses, respectively. They then reason that the "viscous" energy losses correspond to Darcy's law, i.e., in the hydraulic radius theory of permeability to the Carman expression (cf. equation (6.3.3.1)):

$$a = 5\alpha \frac{(1-P)^2}{P^3} \mu S_0^2. \tag{7.2.2.17}$$

The kinetic energy losses would, however, be identical to those sustained by turbulent flow in a cylindrical tube:

$$b = \beta/(2\delta). \tag{7.2.2.18}$$

For a porous medium composed of cylindrical tubes, we have, furthermore:

$$\delta = \frac{4P}{S_0(1-P)}$$

and with the Dupuit-Forchheimer assumption of $q = vP$, we finally obtain

$$|\operatorname{grad} p| = 5\alpha \frac{(1-P)^2}{P^3} \mu S_0^2 q + \frac{\beta}{8} \frac{1-P}{P^3} \rho q^2 S_0. \tag{7.2.2.19}$$

The factors α and β indicate what part of the pressure drop is due to viscous and what part is due to kinetic energy losses. In this instance, they presumably depend on the critical Reynolds number for the system, and as the latter cannot be determined in any universal fashion, the coefficients α and β likewise cannot be determined universally. This seems to subject the present theory to the same limitations as those based upon correlations of the Reynolds number.

An approach similar to Ergun's has been proposed by Cornell and Katz (1953). These authors also start from the assumption of a quadratic equation between grad p and q; they identify the linear term with the Kozeny-type expression, and the quadratic one with a turbulent flow expression. The Kozeny-type equation they use is written in terms of a hydraulic radius instead of S_0, and a tortuosity as measured electrically is introduced. Thus, everything is reduced to measurable quantities—except for the same two factors α and β (Cornell and Katz call them k_1 and k_2) necessary in Ergun's equation.

D. *Drag theory.* Of the theories different from the hydraulic radius theory, the drag theory has also been applied to the problem of high velocity deviation from Darcy's law. Nemenyi (1934) argued qualitatively that the inertia terms which were neglected in Emersleben's theory must effect something like

§7.2   Equations for High Flow Velocities

turbulence in porous media at high flow velocities. Similar investigations have been made by Biesel (1950) and by Shoumatoff (1952).

E. *Statistical theories.* Finally, it remains to apply concepts of statistical hydrodynamics in porous media to the turbulent flow region (Scheidegger, 1955). Following the same procedure as for laminar flow (see sec. 6.5.3) one can introduce a probability-density $v(\boldsymbol{x}, t)$ for a particle to be at the spot $\boldsymbol{x}$ at the time t. As before, this probability density refers to a fictitious ensemble of similar porous media rather than to an actual stochastic flow process. If correlations between adjacent time steps are neglected (as in sec. 6.5.3), the same fundamental distribution for v is obtained as in equation (6.5.3.8):

$$v(\boldsymbol{x}, t) = (4\pi Dt)^{-\frac{3}{2}} \exp\{-(\boldsymbol{x} - \bar{\boldsymbol{x}})^2/(4Dt)\}. \tag{7.2.2.20}$$

The connection between the average displacement $\bar{\boldsymbol{x}}$ and the field of forces (denoted by grad p) is now given, if the simplest case of purely turbulent flow is considered, by:

$$\frac{1}{\tau}|\boldsymbol{\xi}| = -c\sqrt{|\operatorname{grad} p|}\sqrt{\cos\theta}/\sqrt{\rho} \tag{7.2.2.21}$$

where $\boldsymbol{\xi}$ is the displacement during a small time-step τ. Furthermore, c denotes a constant and θ the angle between the vectors $\boldsymbol{\xi}$ and grad p.

Equation (7.2.2.21) is now to be averaged over the ensemble. The average $\bar{\boldsymbol{\xi}}$ will be in the direction $-\operatorname{grad} p$ if the porous medium is assumed to be isotropic, as this is the only distinct direction. Its magnitude will be equal to the average of the component of $\boldsymbol{\xi}$ in the direction of $-\operatorname{grad} p$; i.e., equal to the average of $|\boldsymbol{\xi}|\cos\theta$. Thus we have

$$\bar{\boldsymbol{\xi}}/\tau = -\boldsymbol{n}\,\overline{c\sqrt{|\operatorname{grad} p|}\,\cos^{\frac{3}{2}}\theta}/\sqrt{\rho} \tag{7.2.2.22}$$

or

$$\boldsymbol{V} = -\boldsymbol{n}\,\overline{c\sqrt{|\operatorname{grad} p|}\,\cos^{\frac{3}{2}}\theta}/\sqrt{\rho} \tag{7.2.2.23}$$

where $\boldsymbol{V}$ is now the pore-velocity vector and $\boldsymbol{n}$ a unit vector in direction of grad p. Therefore, the probability distribution can now be expressed as follows:

$$v(\boldsymbol{x}, t) = (4\pi Dt)^{-\frac{3}{2}} \exp\{-[\boldsymbol{x} + t\langle c\sqrt{|\operatorname{grad} p|}\,\cos^{\frac{3}{2}}\theta\rangle \boldsymbol{n}\rho^{\frac{1}{2}}]^2/(4Dt)\}. \tag{7.2.2.24}$$

The average position of the median $\bar{\boldsymbol{x}}$ of this distribution is:

$$\bar{\boldsymbol{x}}(t) = -t\overline{\langle c\sqrt{|\operatorname{grad} p|}\,\cos^{\frac{3}{2}}\theta\rangle}\boldsymbol{n}/\sqrt{\rho} \tag{7.2.2.25}$$

and the average of the square-length deviation is from the general theory of Gauss distributions:

$$\overline{x^2(t)} = 2Dt + \overline{\langle c\sqrt{|\operatorname{grad} p|}\,\cos^{\frac{3}{2}}\theta\rangle^2}t^2/\rho. \tag{7.2.2.26}$$

Thus, the mean pore velocity is:

$$|V| = -\overline{\langle c\sqrt{|\operatorname{grad} p|}\cos^{\frac{3}{2}}\theta\rangle}/\sqrt{\rho}. \qquad (7.2.2.27)$$

The above theory refers to homogeneous forces only. It is, however, easy to effect a generalization to inhomogeneous forces by observing, as was done in section 6.5.3, that the fundamental probability distribution is a solution of a differential equation as follows:

$$\frac{\partial v}{\partial t} = \operatorname{lap}(Dv) + \operatorname{div}\overline{[v\langle c\sqrt{|\operatorname{grad} p|}\cos^{\frac{3}{2}}\theta\rangle\boldsymbol{n}\rho^{-\frac{1}{2}}]}, \qquad (7.2.2.28)$$

which is also valid for inhomogeneous forces. Finally, the continuity condition has to be formulated. Following the same procedure as in section 6.5.3, we end up with the following fundamental equation of turbulent flow:

$$\frac{\partial \rho}{\partial t} = \operatorname{lap}(D\rho) + \operatorname{div}[\rho^{\frac{1}{2}}\boldsymbol{n}\sqrt{|\operatorname{grad} p|}\,\bar{c}\,\overline{\cos^{\frac{3}{2}}\theta}]. \qquad (7.2.2.29)$$

Taking everything together that depends on the porous medium only, this can be written as follows:

$$\frac{\partial \rho}{\partial t} = \operatorname{lap}(D\rho) + \operatorname{div}[m\rho^{\frac{1}{2}}\boldsymbol{n}\sqrt{|\operatorname{grad} p|}] \qquad (7.2.2.30)$$

where m is a constant of the porous medium, and D a certain function that cannot be further evaluated without additional assumptions.

Equation (7.2.2.30) of turbulent flow shows that the latter is composed of two effects; first, one corresponding to the average turbulent flow through a set of small channels; and second, a dispersivity effect. The heuristic equations mentioned earlier do not take the dispersivity into account. It is, however, well known that it does occur.

7.2.3. Solutions of Turbulent Flow Equations.
Compared with the number of solutions of the laminar flow equations available, only a few solutions of the various turbulent flow equations have been investigated.

In order to obtain analytical solutions for turbulent flow through porous media, those equations which one wants to consider as basic, have first to be written in a suitable analytical, i.e., vectorial form. Assuming that the flow is laminar (i.e., of Darcy-type) up to a "critical" Reynolds number, and turbulent (i.e., of Forchheimer-type) above, we can encompass this in a set of equations as follows (Engelund, 1953):

$$-\operatorname{grad} p = F(|q|)\boldsymbol{q}$$

$$F(|q|) = \begin{cases} \mu/k & \text{for} \quad Re < Re_{\text{crit}} \\ a + b|q| & \text{for} \quad Re > Re_{\text{crit}} \end{cases}$$

A suitable modification of (7.2.3.1) can be made to take account of gravity flow simply by introducing instead of p the "hydraulic head" H.

It is, of course, quite hopeless to try to obtain analytical solutions of the system (7.2.3.1) in the general case of three-dimensional, non-steady state flow. Restricting ourselves to the steady-state case in two dimensions, we note that a method of obtaining solutions has been developed by Engelund (1953), as follows.

In two dimensions, the system (7.2.3.1) becomes (Cartesian co-ordinates x, y, z):

$$-\frac{\partial p}{\partial x} = F(q)q_x; \quad -\frac{\partial p}{\partial y} = F(q)q_y. \qquad (7.2.3.2)$$

The equation of continuity is for the steady state:

$$\frac{\partial q_x}{\partial x} + \frac{\partial q_y}{\partial y} = 0. \qquad (7.2.3.3)$$

It is automatically fulfilled if we introduce the stream-function ψ defined by:

$$q_x = -\frac{\partial \psi}{\partial y}; \quad q_y = \frac{\partial \psi}{\partial x}. \qquad (7.2.3.4)$$

From equation (7.2.3.2) it is evident that grad p and q are oppositely directed vectors. Since grad p is perpendicular to the surfaces $p = $ const., the same must hold for the filter velocity vector q, so that the filter stream lines cross the equipressure surfaces at right angles.

It is evident from equation (7.2.3.4) that the vector grad ψ is perpendicular to q, from which it follows that q is tangent to the curves $\psi = $ const. From this one may further conclude that these curves may be interpreted as stream lines. Further it may be concluded from equation (7.2.3.4) that

$$|\text{grad } \psi| = |q| = q. \qquad (7.2.3.5)$$

Hence the stream lines $\psi = $ const., and the contours $p = $ const. form an orthogonal system, just as in laminar flow.

Equations (7.2.3.2) are, however, not linear and they are therefore inexpedient for direct solution, but it is possible to introduce new variables transforming the flow equations into linear equations. Thus, it is convenient to introduce first co-ordinates s and n, denoting the length of arc along curves $\psi = $ const. and $p = $ const., respectively, and next to consider an *infinitesimal element* of flow confined by two neighbouring stream lines and two lines of constant pressure. The equation of continuity can then be written

$$-\frac{1}{\Delta n} \frac{\partial (\Delta n)}{\partial s} = \frac{1}{q} \frac{\partial q}{\partial s}. \qquad (7.2.3.6)$$

Since curl grad $= 0$, the flow equations (7.2.3.2) can be rewritten as follows:

$$\operatorname{curl}\,[F(|\mathbf{q}|)\mathbf{q}] = 0. \tag{7.2.3.7}$$

By means of Stokes' theorem this condition may be expressed by the vanishing of the circulation around any closed curve, i.e., around the element:

$$-\frac{1}{\Delta s}\frac{\partial(\Delta s)}{\partial n} = \frac{1}{Fq}\frac{\partial}{\partial n}(Fq). \tag{7.2.3.8}$$

Engelund then proceeds to introduce as more convenient independent variables the filter velocity q and the angle φ between the velocity vector $\mathbf{q}$ and the x-axis. We obtain for the angle difference between the two stream lines of the element:

$$\frac{\partial \varphi}{\partial n}\Delta n = -\frac{1}{\Delta s}\frac{\partial(\Delta n)}{\partial s}\Delta s. \tag{7.2.3.9}$$

Furthermore, we have

$$\frac{\partial}{\partial s}\left(\frac{\pi}{2}-\varphi\right)\Delta s = -\frac{1}{\Delta n}\frac{\partial(\Delta s)}{\partial n}\Delta n. \tag{7.2.3.10}$$

These expressions reduce to

$$\frac{1}{\Delta n}\frac{\partial(\Delta n)}{\partial s} = -\frac{\partial \varphi}{\partial n}, \quad \text{and} \quad \frac{1}{\Delta s}\frac{\partial(\Delta s)}{\partial n} = \frac{\partial \varphi}{\partial s}. \tag{7.2.3.11}$$

Substitution into equations (7.2.3.6) and (7.2.3.8) yields

$$\frac{\partial \varphi}{\partial n} = \frac{1}{q}\frac{\partial q}{\partial s}, \tag{7.2.3.12}$$

$$-\frac{\partial \varphi}{\partial s} = \frac{1}{qF}\frac{\partial}{\partial n}(qF) = \left(\frac{1}{q}+\frac{F'}{F}\right)\frac{\partial q}{\partial n}, \tag{7.2.3.13}$$

where F' denotes the derivative dF/dq.

The quantities p and ψ can be introduced into these equations by substitution of $\partial \varphi/\partial n = \partial \varphi/\partial \psi \cdot d\psi/dn = q\partial \varphi/\partial \psi$, etc., and thus become

$$\frac{\partial \varphi}{\partial \psi} = -\frac{F}{q}\frac{\partial q}{\partial p}, \tag{7.2.3.14}$$

$$\frac{\partial \varphi}{\partial p} = \frac{1}{F}\left(\frac{1}{q}+\frac{F'}{F}\right)\frac{\partial q}{\partial \psi}. \tag{7.2.3.15}$$

These equations can be solved for q and φ as functions of p and ψ. We may, however, equally well express the functions p and ψ in terms of q and φ, and substitute this into the flow equations. The latter then become

$$\frac{\partial \psi}{\partial \varphi} = -\frac{q}{F}\frac{\partial p}{\partial q}, \qquad (7.2.3.16)$$

$$\frac{\partial \psi}{\partial q} = \frac{1}{F}\left(\frac{1}{q} + \frac{F'}{F}\right)\frac{\partial p}{\partial \varphi}. \qquad (7.2.3.17)$$

Furthermore, ψ can be eliminated from these equations by appropriate differentiation, which leads to

$$\frac{\partial}{\partial q}\left(\frac{q}{F}\frac{\partial p}{\partial q}\right) + \frac{1}{F}\left(\frac{1}{q} + \frac{F'}{F}\right)\frac{\partial^2 p}{\partial \varphi^2} = 0, \qquad (7.2.3.18)$$

and thus one arrives at a single partial differential equation for p which describes laminar as well as turbulent steady state flow in porous media. Engelund (1953), who developed this equation, showed how it can be solved for particular cases.

In the extreme case for high Reynolds numbers, the quadratic term in the equation (7.2.3.1) is preponderant (purely turbulent flow), and (7.2.3.18) can be reduced to

$$\frac{\partial^2 p}{\partial q^2} + \frac{2}{q^2}\frac{\partial^2 p}{\partial \varphi^2} = 0. \qquad (7.2.3.19)$$

Again, Engelund lists a set of particular solutions. As an illustrative example, we reproduce here Engelund's solution for the symmetrical radial flow to a single well.

On account of symmetry, p is independent of φ and (7.2.3.19) becomes

$$\frac{\partial^2 p}{\partial q^2} = 0, \qquad (7.2.3.20)$$

from which

$$p = c_1 q + c_2. \qquad (7.2.3.21)$$

Thus, p is in this case a linear function of q. To see how p depends on the distance r from the axis of the well, one puts

$$dp/dr = bq^2 = c_1 dq/dr. \qquad (7.2.3.22)$$

The solution of this equation is

$$q = -\frac{c_1}{br} \quad \text{or} \quad c_1 = -2\pi q r \frac{b}{2\pi} = Q\frac{b}{2\pi} \qquad (7.2.3.23)$$

where Q denotes the discharge per unit length of the well. Thus, one has

$$p = \frac{bQ}{2\pi} q + c_2 = \frac{Q^2 b}{4\pi^2 r} + c_2. \qquad (7.2.3.24)$$

Solutions of the more general flow equation of Engelund (7.2.3.18) are much more tedious to obtain. Engelund lists a variety of methods and applies them to a series of special cases which are of interest in the theory of groundwater flow into wells, drainage tubes, etc. The reader is referred for the details to the cited paper of Engelund (1953).

7.3. Equations Accounting for Molecular Effects

7.3.1. Gas Slippage and Molecular Streaming. The existence of molecular phenomena that can effect a limitation of Darcy's law, has already been discussed in section 7.1.2. From such experiments, Sameshima (1926) deduced an experimental flow equation for gases through porous plates. He found

$$1/q = C \mu^n M^{(1-n)/2} \qquad (7.3.1.1)$$

where M is the molecular weight, and C and n are constants independent of the gas but dependent on the porous medium.

Adzumi (1937b) seems to have been the first to give an explanation of gas flow equations of the form of (7.3.1.1) in terms of molecular slip. He constructed a theoretical model of porous media. The porous medium is represented by a bundle of parallel capillaries, each of which is made up of a series of short capillaries of various diameters. Using a modification of Knudsen's law [Adzumi (1937a); see equation (2.3.4)] for the (slip) flow through *one* capillary, Adzumi arrived at an equation for the flow of a gas through the porous medium which may be written as follows:

$$q = \frac{\pi \Delta p E}{8 A \mu} + \varepsilon \frac{4}{3} \sqrt{2\pi} \sqrt{\frac{R \mathfrak{T}}{M}} \frac{F}{Ap} \Delta p \qquad (7.3.1.2)$$

where A equals the cross-sectional area of the porous medium; ε is the Adzumi constant ($= 0.9$, cf. equation 2.3.4); $E = n R_0^4 / h$; and $F = n R_0^3 / h$; with $n =$ number of pores in A; $R_0 =$ average radius of the pores; $h =$ thickness of the porous medium. R, M, $\mathfrak{T}$ are gas constant, molecular weight, and absolute temperature respectively.

The Adzumi equation explains the occurrence of a "slip term" in the flow equation by means of a simple capillaric model of the porous medium. It is, of course, quite hopeless to expect that E and F can ever be calculated for an actual porous medium; they therefore will have to be measured.

The train of thought originated by Adzumi has been taken up by many authors with the intention of making refinements on Adzumi's model, mainly by starting out with the Kozeny equation for the nonmolecular part of the flow, instead of with a model. This has been done by Arnell (1946, 1947; Arnell and Henneberry, 1948), and by Carman (1947, 1950; Carman and Arnell, 1948). Similar investigations have been undertaken by Hodgins, Flood and Dacey (1946), Holmes (1946), Keyes (1946), Rigden (1946, 1947), Brown et al. (1946), Barrer (1948), and by Wilson et al. (1951).The following exposition of the available modifications of the Adzumi theory is based upon a review article by Carman and Arnell (1948).

Thus, Rigden (1947) applied the same considerations as used to deduce the Kozeny equation to the Poiseuille equation corrected for "slip" and arrived at an equation having the form

$$q = \operatorname{grad} p \frac{Pc}{\mu} \left[\left\{ \frac{P}{S_0(1-P)} \right\}^2 + \frac{2Px\lambda}{S_0(1-P)} \right] \qquad (7.3.1.3)$$

where c is a Kozeny constant approximately equal to $\frac{1}{5}$, x is a constant having the value of 0.874 and λ is the mean free path of the gas molecules. However, the slip correction factor used by Rigden was in the form valid for circular capillaries, whereas, for other media, the correction should be larger. Thus, the use of a slightly modified equation worked out by Lea and Nurse (1947) is preferable:

$$q = \operatorname{grad} p \frac{Pc}{\mu} \left[\left\{ \frac{P}{S_0(1-P)} \right\}^2 + \left(\frac{2}{f} - 1 \right) \sqrt{\frac{\pi}{2} \frac{R\mathfrak{T}}{M}} \frac{\theta \mu P}{S_0(1-P)p} \right]. \qquad (7.3.1.4)$$

where f = fraction of molecules evaporated from the surface, or $(1 - f)$ = the fraction of molecules undergoing elastic collisions with the surface, and p = mean pressure of the gas. Furthermore, θ is a shape factor which generally lies between 2 and 3. Finally, Arnell (1946), by analogy with Knudsen's flow law through capillaries, arrived at the following equation

$$q = \operatorname{grad} p \frac{Pc}{\mu} \left[\left\{ \frac{P}{S_0(1-P)} \right\}^2 + \frac{8}{3} \sqrt{\frac{2R\mathfrak{T}}{\pi M}} \frac{\mu C\alpha}{cS_0(1-P)p} \right] \qquad (7.3.1.5)$$

where C is a variable factor having a value of approximately 0.9 and α is given by the equation $\log_{10} \alpha = 1.41P - 1.40$. Carman's and Arnell's equations are identical if $f = 0.79$. Experimental checks of these equations seem to give plausible agreement, but the tests and the equations are open to the same criticisms as is the Kozeny equation.

A number of years after Adzumi, Klinkenberg (1941) also came up with an explanation of the observed gas flow "anomalies," apparently without knowing about Adzumi's work. Klinkenberg constructed a very simple capillaric model of porous media and applied Warburg's slip theory to each capillary. In this way, he arrived at a flow equation for gases in porous media which allows for slip effects. The Klinkenberg equation has been tested by Grunberg and Nissan (1943), Calhoun (1946), Calhoun and Yuster (1946), Heid et al. (1950) and Collins and Crawford (1953). The Klinkenberg theory has been compared to that of Adzumi by Rose (1948) who duly showed that the two theories are identical: Adzumi studied the dependence of the filter velocity q upon the average pressure p_m, whereas Klinkenberg introduced the "superficial" gas permeability k_a, defined by

$$q = -\frac{k_a}{\mu} \operatorname{grad} p$$

and investigated its dependence on the *reciprocal* mean pressure. Klinkenberg found that this dependence should be linear if slip is the correct explanation of gas flow:

$$k_a = b + m/p_m. \tag{7.3.1.6}$$

This last equation is, in fact, an equivalent expression of Adzumi's statements. The quantities b and m are constants.

There have also been attempts at explanations of anomalies in gas flow which fall somewhat out of line with those discussed above. Deryagin, Fridlyand and Krӯlova (1948) used a modification of Knudsen's flow law for capillaries and applied it to capillaric models of porous media. Iberall (1950) extended the drag theory of permeability to include slip flow. In this fashion, he arrived at an equation for gas flow of the following type

$$q = c(1 + bp_0/p_m) \operatorname{grad} p \tag{7.3.1.7}$$

where p_0 is the reference pressure at which q is measured, p_m is the mean pressure, and b and c are constants. The last formula is the same as that given by Klinkenberg. Iberall claims fair agreement of the theoretically computed constants b, c with experiments.

Finally, one can also apply statistical considerations to the molecular flow problem in porous media (Scheidegger, 1955). Following procedure analogous to that outlined in section 6.5.3 for laminar flow or section 7.2.2 for turbulent flow, one ends up, in the present case of molecular flow, with the following equation:

$$\frac{\partial \rho}{\partial t} = \operatorname{lap}(D\rho) + \operatorname{div}\left[(c_1 + c_2/p)\rho \operatorname{grad} p\right] \tag{7.3.1.8}$$

where D is a diffusivity function (cf. 7.2.2.26) and c_1 and c_2 are constants. This last equation demonstrates that flow through porous media in the molecular régime can indeed be described by the flow through an assemblage of capillaries in each of which Knudsen flow occurs, but there is, in addition, a diffusivity effect which originates from the statistical considerations.

Turning now to solutions of molecular flow laws for particular cases, we note that, in order to achieve them, we have to choose a particular law and write it in differential form. Using the representation of Klinkenberg (7.3.1.6), we obtain, in conjunction with (i) Darcy's law, (ii) the pressure-density relation for an ideal gas, and (iii) the continuity equation:

$$\operatorname{div}\left[\left(p + \frac{m}{b}\right)\operatorname{grad} p\right] = \frac{P\mu}{b}\frac{\partial p}{\partial t}. \qquad (7.3.1.9)$$

Equation (7.3.1.9) is a nonlinear differential equation and therefore difficult to treat analytically. Thus, Aronofsky (1954) developed a numerical method which makes the problem accessible for high speed computing machines. Introducing the new variable W ("equivalent pressure")

$$W = p + m/b, \qquad (7.3.1.10)$$

and substituting into equation (7.3.1.9), we obtain

$$\operatorname{div}(W \operatorname{grad} W) = \frac{P\mu}{b}\frac{\partial W}{\partial t}. \qquad (7.3.1.11)$$

The problem of finding particular flow patterns of molecular streaming in porous media is therefore reduced to finding solutions, suitable for prescribed boundary conditions, of equations (7.3.1.11) which is still a nonlinear differential equation. Aronofsky's (1954) numerical method consists in replacing the differential equation (7.3.1.11) by a difference equation (applicable to the one-dimensional case):

$$\left(\frac{W}{W_m}\right)_{x,\,t+\Delta t} = \frac{1}{4}\left[\left(\frac{W}{W_m}\right)^2_{x+\Delta x,\,t} + \left(\frac{W}{W_m}\right)^2_{x-\Delta x,\,t} - 2\left(\frac{W}{W_m}\right)^2_{x,\,t}\right] + \left(\frac{W}{W_m}\right)_{x,\,t} \qquad (7.3.1.12)$$

where the relation between time step and co-ordinate step has been chosen as follows (W_m is a constant):

$$\frac{1}{2}\frac{\Delta t W_m b}{(\Delta x)^2 P\mu} = \frac{1}{4}. \qquad (7.3.1.13)$$

Aronofsky (1954) lists several cases for which he has obtained solutions; as an example, we present one of them here in some detail.

Consider a finite tube of porous medium charged with an initial gas pressure p_m (equivalent pressure W_m). The pressure at one end of the tube ($x = 0$) is suddenly lowered to a constant value p_0 (equivalent pressure W_0); the other end of the tube is sealed so that no flow occurs across the plane $x = L$.

By iterating equation (7.3.1.12) numerically on a computing machine, it was found that the pressure declines at the sealed end of the tube ($x = L$) are as shown in figure 19. In this figure, the equivalent pressure ratio at the sealed end

$$EPR \equiv (W(L, t) - W_m)/(W_0 - W_m) \qquad (7.3.1.14)$$

is plotted as a function of a dimensionless time parameter

$$\tau_w = bW_m t/(P\mu L^2). \qquad (7.3.1.15)$$

A family of curves is shown for various values of $H = W_0/W_m \leq 1$. The spread of the curves can be taken as an indication of the nonlinearity of the basic differential equation (7.3.1.11). The solution for molecular streaming approaches

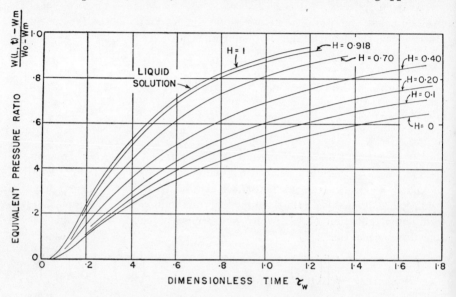

Figure 19. Pressure draw-down curves at the sealed end of a tube of porous medium charged with gas when the other end is suddenly opened to a constant pressure. (After Aronofsky, 1954.)

the laminar flow solution in the limit of $H \to 1$. The laminar flow solution can be obtained by direct integration.

7.3.2. Adsorption and Diffusion. The permeability of a medium to a fluid under the action of a pressure difference may arise in still other ways than those discussed heretofore in this study: (i) the fluid may dissolve and be transported by diffusion along a concentration gradient produced by the pressure difference; (ii) the fluid may be adsorbed at the internal walls of the pores and be transported by diffusion along a concentration gradient. In either case, the appearance of the phenomenon is that of diffusion of the fluid through the solid and is described by a diffusivity equation. In the present study, we shall not concern ourselves with the phenomenon of actual dissolution of the fluid in the solid as this is a chemical process, but only with the second possibility which is characteristic for the motion of fluids through microporous substances such as activated carbon, etc.

§7.3 Equations Accounting for Molecular Effects

The flow of a fluid through a porous medium, if adsorption occurs, has been discussed by Bull and Wronsky (1937) and later by Wicke (1938, 1939a, 1939b; Wicke and Kallenbach, 1941) who established the phenomenon as one of surface flow in the adsorbed layers. Other early studies are due to Sandera and Mirčev (1938), Barrer (1939, 1941,) Penman (1940a, 1940b), Glückauf (1944) and Cassie (1945). Wentworth (1944) investigated deviations from Darcy's law in assemblages of thin cracks (such as lava) and showed that

$$q \sim (\text{grad } p)^{1/N} \tag{7.3.2.1}$$

with $N < 1$. In this particular case, the rates were more than expected from Darcy's law as the molecules of the fluid within the sorbed layers are generally assumed as more mobile. Further investigations on this subject have been made by Briant (1945), Takagi (1947), Wicke and Voigt (1947), Tomlinson and Flood (1948), Barrer (1949), Berg et al. (1949), Happel (1949), and Friedel (1949). In a series of papers, Carman and co-workers (Carman 1949, 1952; Carman and Malherbe 1950a; Carman and Raal, 1951) gave a comprehensive exposition of experiments and theories in connection with flow through adsorbents.

Thus, Carman and Malherbe (1950a) approach surface flow as an example of diffusion along a concentration gradient. Following their exposition, we assume adsorption equilibrium at the end pressures, p_1 and p_2, of a cylindrical piece of porous medium with parallel end faces. The quantities adsorbed are then y_1 and y_2 millimoles per gram of adsorbent where y_1 and y_2 are obtained from the adsorption isotherm for the temperature of the porous medium. As the weight of adsorbent per unit volume of porous medium is $\rho_p(1 - P)$ (ρ_p being the density of the solid material), the concentration gradient is equal to $\rho_p(1 - P)\Delta y/\Delta L$ where $\Delta y = y_1 - y_2$, and ΔL is the distance of the two faces of the porous medium from each other. Then, if the flow rate by surface diffusion is Q millimoles per second,

$$Q\Delta L/A = K\rho_p(1 - P)\Delta y, \tag{7.3.2.2}$$

where K is a diffusion coefficient with dimensions cm^2/sec and A signifies the area of the faces of the porous medium. The concentration gradient is arrived at by assuming equilibrium with the gaseous phase at the two faces. When there is no surface flow, permeability is governed by equation (7.3.1.4). If now it is assumed that surface flow and gas flow take place independently and in parallel, the total flow is obtained by adding the surface term to (7.3.1.4). With properly chosen constants (K, K', K''), one obtains:

$$\frac{Q\Delta L}{A\Delta p} = K\rho_p(1 - P)\frac{\Delta y}{\Delta p} + \frac{K'P^2}{(1 - P)S_0\sqrt{MR\mathfrak{T}}} + \frac{K''pP^3}{\mu S_0^2 R\mathfrak{T}(1 - P)^2}. \tag{7.3.2.3}$$

The last two terms are the slip flow and viscous flow terms, respectively, for flow in the gas phase.

The assumption of independent surface, viscous, and slip flow cannot be accurate, since gas flow must be blocked to some extent by the adsorbed films. It turns out that, in the majority of cases, the flow in an adsorbed layer can indeed be expressed satisfactorily as diffusion along a concentration gradient, but if capillary condensation plays a major role, this is certainly not true. A different mechanism, therefore, must take over in the latter case.

The effect of capillary condensation can be estimated as follows (Carman, 1952). Consider a cylindrical porous medium across which a condensable vapour is flowing under a pressure difference $\Delta p = p_2 - p_1$ across the medium, which is steadily maintained. As an approximation, one might assume that the condensate behaves as if it were bulk liquid in viscous flow across the porous medium. The driving force producing the flow, however, would then not be the maintained pressure difference in the gas, but a very much larger difference of capillary pressure resulting from this maintained pressure difference, presumably owing to action of surface tension at curved menisci. This pressure difference, in turn, can be estimated as follows.

Capillary condensation must be produced by a capillary potential, expressible as negative pressure, or tension, $\bar{P}$, which must be applied to bulk liquid to reduce its vapour pressure from p_0 to that of the condensate, p. It is given by

$$\bar{P} = \frac{\rho_L R \mathfrak{T}}{M} \log \text{nat} \frac{p_0}{p} \tag{7.3.2.4}$$

where ρ_L = liquid density at temperature $\mathfrak{T}$. It follows that

$$\Delta \bar{P} = \frac{\rho_L R \mathfrak{T}}{M} \log \text{nat} \frac{p_1}{p_2}. \tag{7.3.2.5}$$

As long as Δp is less than half p_1 or p_2 (the end pressures), this is sufficiently approximated by

$$\Delta \bar{P} = \frac{\rho_L R \mathfrak{T}}{M} \frac{\Delta p}{p} \tag{7.3.2.6}$$

where $p = \frac{1}{2}(p_1 + p_2)$.

A simpler derivation of equation (7.3.2.11) can be obtained by writing down the true thermodynamic relationship between $\bar{P}$ and p:

$$\frac{dp}{d\bar{P}} = \frac{\rho_g}{\rho_L}, \tag{7.3.2.7}$$

where ρ_g is the gaseous density at pressure p. Equation (7.3.2.6) then follows by substituting Δp and $\Delta \bar{P}$ for differentials and assuming that the gas is ideal.

It follows that the value of $\Delta \bar{P}$ created by Δp across the sample is some hundreds of times larger than Δp.

Now, though $\bar{P}$ represents a tension which must be applied to bulk liquid to produce a vapour pressure, p, it does not follow that it actually exists in adsorbed films, since these are in a state different from bulk liquid. In the case of capillary condensate, however, it is usual to assume (in approximation) that it possesses the properties of bulk liquid, including the same surface tension, and that reduction of vapour pressure is wholly due to the tension, $\bar{P}$, which is produced by surface tension at curved menisci. It follows that a difference, $\Delta \bar{P}$, can be regarded as a *real* difference of pressure in the capillary condensate at the two ends of the sample. If we further assume that the viscosity is also that of bulk liquid, it is only a short step to envisage flow of capillary condensate as viscous flow through the pore space under the pressure difference, $\Delta \bar{P}$, as outlined above.

Thus, the flow of capillary condensate can indeed be regarded as actual viscous flow under the pressure difference $\Delta \bar{P}$. Carman (1952) substantiated this by a series of experiments.

It is thus seen that flow in adsorbents can be visualized correctly, at least qualitatively, by assuming either capillary condensate flow or else adsorbate flow along a concentration gradient. These two assumptions constitute two opposite and extreme cases and, in practice, one will find oneself somewhere in between the two. The theories given by Carman for these extremes are only approximate ones, but they do give at least a qualitative explanation for the observed phenomena. A gratifying aspect of these two theories is that they do not make use of any hydraulic radius models as far as the respective surface flow terms are concerned. As was shown above, the derivation of the surface flow terms is solely based upon thermodynamic reasoning and is therefore much more trustworthy than the derivation of the viscous and Knudsen terms also contained in the corresponding final equations.

The phenomenon of surface flow in porous media has received further study by Barrer and co-workers (Barrer and Grove, 1951a, 1951b; Barrer and Barrie, 1952). These investigations are only concerned with the sorbed (not the capillary-condensate) state of the fluid. By measuring the time-lag in setting up the steady state flow, it was possible to measure the diffusion coefficient of gases and to obtain a mean pore radius and an internal surface of the porous medium.

Other investigations on the flow of fluids in adsorbents have been made by Flood *et al.* (1952a, 1952b, 1952c), who developed a simple flow equation for the correlation of their experimental data, and by Jones (1951, 1952). The latter author derived theoretical equations for the surface flow of an ideal two-dimensional gas through a porous material, the latter being treated as a bundle of capillaries. These equations explain earlier results by Tomlinson and Flood (1948) if the conditions are such that nearly a monolayer is sorbed.

Finally, there have been some recent experimental studies of the subject. Such investigations have been undertaken by Pfalzner (1950), Danckwertz (1953), Hoogschagen (1953), and Wicke and Trawinski (1953).

In view of all these results, the necessity of adding a surface-flow term to the Darcy and Knudsen terms in flow through adsorbent porous media, seems definitely established.

7.4. Applications of General Flow Equations

7.4.1. Permeability Corrections.

The phenomena described by general flow equations may have the effect that the superficial permeability of a porous medium varies under different conditions. Under superficial permeability is understood the expression corresponding to k in Darcy's law if the latter *were* valid.

Thus, if "permeability" measurements are made under conditions different from those to which they are to be applied, "corrections" have to be made. This is particularly important if one is interested in the permeability of a porous medium to liquid, but the measurements are made employing a gas. A correction is best effected by using the representation of the slip effect given by Klinkenberg (1941; see sec. 7.3.1), where it was shown that the superficial gas permeability is a linear function of the inverse mean pressure of the gas. Since the cause of this is supposed to be gas slippage which in turn increases with increasing free path length, the effect should be nullified if the free path length is zero, i.e., if the mean pressure is infinite. Gas and liquid permeability should therefore be identical if the gas pressure is infinite. Using the equation of Klinkenberg, it is simple to extrapolate from two measurements at finite mean pressures to the infinite one. The routine application of this "Klinkenberg correction" of gas permeabilities to obtain liquid ones, has been described, e.g., by Krutter and Day (1941).

7.4.2. Structure Determinations of Porous Media.

As with the Kozeny equation, those types of general flow equations which contain the specific surface S or S_0 can be used for structure determinations of porous media.

Of the turbulent flow equations, Ergun (1951, 1952b) used his equation for the determination of particle density and of geometric surface area of crushed porous solids.

Most applications, however, use the slip flow equations in one form or another. Some of these applications have already been mentioned in section 6, others are due to Lochmann (1940), Arnell (1946, 1947, 1949), Arnell and Henneberry (1948), Deryagin (1946), Holmes (1946), Lea and Nurse (1947), Rigden (1947), Carman and Arnell (1948), Carman and Malherbe (1950b), Schwertz (1949), Kraus and Thiem (1950), and Wilson et al. (1951).

Most of these applications are structure determinations of industrial powders. Owing to the fact that these powders are rather uniform to begin with, the success of the method does not necessarily imply the quantitative correctness of the formulas employed.

PART VIII

MULTIPLE PHASE FLOW

8.1. General Remarks

Contributions to the study of multiple phase flow are much fewer in number than those concerning single phase flow. This is owing to the somewhat restricted applicability of multiple phase flow, as compared with single phase flow, to practical cases. Nevertheless, in those fields where multiple phase flow plays any role at all, it has a very important one. Contributions to the subject have been noted from the following fields:

(i) from the oil industry: the simultaneous flow of oil, water, and gas in porous rock strata is important in connection with the production of oil from oil fields;
(ii) from the study of countercurrent towers in chemical engineering;
(iii) from soil science: the flow of moisture in unsaturated soils (i.e., the simultaneous flow of water and air) is of importance;
(iv) from groundwater hydrology: much attention has been paid to the encroachment of salt water into fresh water in reservoirs beneath oceanic islands.

The study of multiple phase flow in porous media could be split into sections similar to those for single phase flow. Thus one could characterize the various cases by the régime of flow prevalent in each; viz., whether the flow is laminar, turbulent, molecular, etc.

However, it turns out that a much more important distinction than that of flow régime is one concerning the fluids: viz., whether the latter are miscible or immiscible. It also turns out that the study of immiscible fluids is much simpler than that of miscible fluids. Thus most of the literature concerned with multiple flow relates to immiscible fluids, and relatively little to miscible fluids.

One should therefore characterize the various aspects of multiple phase flow by the flow régime as well as by whether the fluids are miscible or not. However, since only a little is known about miscible displacement in porous media, most of Part VIII will be devoted to immiscible fluids. Sections 8.2–8.6 will be concerned with the various flow régimes of immiscible fluids, and section 8.7 will relate what is known about miscible displacement.

It may be interesting to observe that in the oil industry, in chemical engineering, and in soil science, the chief concern is with immiscible displacement, whereas the hydrological problem of displacement of fresh water by sea water beneath an oceanic island constitutes a classical example of miscible displacement. Lately the oil industry has also become concerned with miscible displacement in connection with the possibility of flushing oil by solvents from oil reservoirs. It is therefore to be expected that advances will be made in the near future in the study of miscible displacement, whereas the study of immiscible fluids in porous media is, in its physical principles at least, more or less complete.

There seem to be no monograph of multiple phase flow in porous media worthy of that name. Most text-books on oil reservoir engineering contain some concepts (Spencer, 1949; Muskat, 1949; Pirson, 1950; Calhoun, 1953) and so does a general article by Johnston and van Wingen (1945). These references, however, do not systematically outline the physical principles of multiple phase flow; for such an exposition one has to go back to the original literature.

8.2. Laminar Flow of Immiscible Fluids

8.2.1. Qualitative Investigations.
In line with single phase flow experiments intended to test Darcy's law, similar experiments have been performed using two immiscible phases. In the early days it was assumed that Darcy's law should be valid for any fluid and thus also for a mixture of two immiscible fluids. Studies to test this assumption have been performed by Cloud (1930), Gardescu (1930), Versluys (1931), Bartell and Miller (1932), Uren and Bradshaw (1932), Garrison (1934), Uchida (1937), Uren and Domerecq (1937), Plummer et al. (1937), Christiansen (1944), Pillsbury and Appleman (1945), Fletcher (1949), and Pirverdyan (1952). Experimental studies in this series soon rendered it evident that the presence of a second phase not only makes the permeability to the first much lower, but also greatly decreases the permeability to the mixture. Many qualitative arguments to explain this fact have been put forth. Gardescu (1930), for instance, ascribed it to an effect analogous to one noticed many years ago by Jamin (1860) in small tubes. Jamin observed that in a capillary tube containing a large number of detached drops of liquid interspersed with gas, a large difference in pressure between the end of the tube without any appreciable movement of the drops along the tube may occur. The effect is ascribed to hysteresis of contact angle.

The Jamin effect gives a qualitative explanation of the observed "threshold" pressures necessary before flow can occur. However, in fact, double phase flow in porous media probably does not occur in the manner observed by Jamin, but rather in the funicular channels postulated by Versluys (see Part III). The latter author also applied the concepts of funicular and pendular saturation regions to flow problems (Versluys, 1931) by giving a series of qualitative microscopic arguments for the explanation of the observed qualitative results.

8.2.2. Darcy's Law.
The above qualitative investigations are not sufficient to provide a quantitative description of double phase flow in porous media.

§8.2 LAMINAR FLOW OF IMMISCIBLE FLUIDS

It is therefore necessary to extend the earlier investigations to obtain a quantitative description.

As a working assumption, one might be inclined to postulate that Darcy's law should be valid for *each* flowing phase. In this manner, it is possible to formulate a complete hydrodynamics for multiple phase flow.

In single phase flow, Darcy's law can be formulated as follows (cf. 4.3.1.2):

$$q = - (k/\mu) (\text{grad } p - g\rho) \qquad (8.2.2.1)$$

where q is the seepage velocity vector, k the permeability, μ the viscosity, p the pressure, g the gravity vector and ρ the density of the fluid. The law of Darcy is not sufficient to determine a given flow problem; in addition, there is the continuity equation

$$- P \frac{\partial \rho}{\partial t} = \text{div}(\rho q) \qquad (8.2.2.2)$$

and a relationship between ρ and p:

$$\rho = \rho(p). \qquad (8.2.2.3)$$

These three equations are the necessary and sufficient conditions, in conjunction with boundary conditions, to determine any given single phase flow problem.

The above conditions can be extended for double phase flow in an obvious way. Denoting the two phases by the subscripts 1 and 2 respectively, we have

$$q_1 = - k \frac{k_1}{\mu_1} (\text{grad } p_1 - g\rho_1) \qquad (8.2.2.4\text{a})$$

$$q_2 = - k \frac{k_2}{\mu_2} (\text{grad } p_2 - g\rho_2) \qquad (8.2.2.4\text{b})$$

$$P \frac{\partial(\rho_1 s_1)}{\partial t} = - \text{div }(\rho_1 q_1) \qquad (8.2.2.5\text{a})$$

$$P \frac{\partial(\rho_2 s_2)}{\partial t} = - \text{div }(\rho_2 q_2) \qquad (8.2.2.5\text{b})$$

$$s_1 + s_2 = 1 \qquad (8.2.2.5\text{c})$$

$$\rho_1 = \rho_1(p_1) \qquad (8.2.2.6\text{a})$$

$$\rho_2 = \rho_2(p_2) \qquad (8.2.2.6\text{b})$$

$$p_2 - p_1 = p_c(s_1) \qquad (8.2.2.7)$$

Here, the symbols are defined as follows: k = total permeability, $k_{1,\,2}$ = relative permeability (in fractions of the total permeability), s = saturation, P = porosity, and p_c = capillary pressure. The above equations *define* the notion of "relative permeability." Sometimes the relative permeability is also given in per cent of the total permeability, rather than as a fraction thereof.

The above extension of Darcy's law to double phase flow has been suggested by Muskat and co-workers (Muskat and Meres, 1936; Muskat et al., 1937). The capillary pressure term seems to have been first introduced by Leverett (1941) who also postulated the validity of an obvious extension of the system of equations (8.2.2.4) to (8.2.2.7) for three phases (Leverett and Lewis, 1941). Further discussions of the principles of the multiple phase flow equations have also been given by Moore (1938), Childs (1945), Muskat (1949), Hubbert (1950), and Rose (1954). Fatt (1953) showed that the compressibility of the porous medium could also be taken into account in the equations for multiple phase flow in a manner similar to that for single phase flow.

The above deduction of Darcy's law for multiple phase flow is, so far, only a theoretical speculation. It is, of course, always possible to *define* relative permeability as in equations (8.3.3.4) to (8.2.2.7), but these equations make sense only if that relative permeability is independent of pressure and velocity, i.e., if it is a function of saturation only. From a great number of investigations it appears that the relative permeabilities to *immiscible* fluids are indeed *within limits* such functions of saturation only. Experiments to substantiate this have been performed, first by the originators of the equations (8.2.2.4 ff.), and also by Wyckoff and Botset (1936), Hassler *et al.* (1936), Reid and Huntington (1938), Botset (1940), Krutter (1941), Averyanov (1949), Kinney and Nielsen (1949, 1950), and Richardson *et al.* (1952). Terwilliger and Yuster (1946, 1947) made a careful study of the possible effects of chemical agents. They, and later Calhoun (1951), found that the chemical composition of the fluids does not matter much and that the relative permeability functions are approximately the same for any "wetting-nonwetting" system. It should be noted, however, that accurate experiments show that this is, as indicated, only *approximately* true. The generalized Darcy law for multiple phase flow has, as ought to be expected, limitations. Some of these limitations are due to the same causes as the limitations of the Darcy law in single phase flow (such as adsorption, molecular slip, turbulence, etc.) and will be dealt with later. However, the assumption that the relative permeability is a function of saturation only, can possibly be only an approximation. Owing to the occurrence of hysteresis in wettability, the same phenomenon must occur in relative permeability to a greater or less extent.

The generalization of Darcy's law presented here corresponds to the force potential formulation of the law for single phase flow. The fact that this generalization seems to lead to an approximately correct description of multiple phase flow, is a strong argument for preferring this formulation of Darcy's law to that based upon the velocity potential.

The system of differential equations representing multiple phase flow can be slightly generalized if further parameters are introduced which may be associated with the two phases. Such parameters have to fulfil conservation laws, which will provide the additional equations necessary to make the problem determined. In practice, such a parameter will usually be an energy quantity (such as temperature), because energy satisfies a conservation law. However, it is conceivable that it could be some other constant of the motion.

8.2.3. Measurement of Relative Permeability. The measurement of relative permeability consists, essentially, of the determination of two flow rates under a given pressure drop and the determination of saturation. Seven methods are commonly applied, termed (i) the Pennsylvania State method; (ii) the single sample dynamic method; (iii) the gas drive method; (iv) the stationary liquid method; (v) the Hassler method; (vi) the Hafford technique; and (vii) the dispersed feed method. General discussions and reviews of these methods have been given by Brownscombe et al. (1949, 1950, 1950b), Rose (1951a, 1951b), Osoba et al. (1951), and Richardson et al. (1952). A compilation of the virtues of the seven methods is given in table V. The description of the seven methods given in this section is based very closely upon the reviews of Osoba et al. (1951) and of Richardson et al. (1952).

TABLE V

A Comparison of Various Techniques for Relative Permeability Determination

Method	Reliability of results	Speed Hrs./Sample (permeability 10^{-9} cm.^{2})	Simplicity of operation	Remarks
Penn state	Excellent	8	Complicated	Uses three samples
Hassler	Excellent	40	Very complicated	Requires pressure gauges of very low displacement volume
Single sample dynamic	Questionable for short samples	6	Simple	For short samples the relative permeability to wetting phase is too high
Stationary liquid	Questionable at low gas saturations	4	Simple	Applicable only to measurement of relative permeability to gas
Gas drive	Good	2	Very simple	Can be operated with minimum amount of training and requires a minimum amount of equipment
Hafford	Excellent	7	Simple	Preferable to dispersed feed
Dispersed feed	Excellent	7	Simple	

The *Pennsylvania State method* (after Osoba et al., 1951) was developed by Yuster and co-workers (Morse et al., 1947a, 1947b; Henderson and Yuster 1948a, 1948b). It was modified by Caudle et al. (1951). The sample to be tested is mounted between two samples of similar material, the three being in capillary contact. This minimizes boundary effects and also effects a good mixing of the two phases before proceeding to the sample. The sample is first saturated with one phase and then this phase is allowed to flow through it at a rate that gives a predetermined pressure drop. The second phase (a gas) is then allowed to flow at a very low rate. After the system reaches equilibrium, the test sample is removed quickly, and the saturation is determined by weighing. The samples are then re-assembled, and flow of both phases is resumed at the rates that had existed previously. The rate of flow of the first phase is then decreased slightly, and the rate of flow of the gas is increased simultaneously to maintain the pressure drop across the sample at its previous value. After equilibrium has again been attained, the test section is removed and the saturation is determined. This step-wise procedure is repeated until a complete relative permeability curve is obtained at incremental changes in the fluid saturation. The Pennsylvania State method is very accurate, but rather complicated to carry out.

The *single sample dynamic method* (after Osoba et al., 1951) uses equipment similar to that employed in the Pennsylvania State technique, except that a single sample, mounted in Lucite, is used. The two phases are introduced directly into the latter from a feed head. In making determinations by this method, the sample is initially saturated completely with the first phase. A rate of flow of the latter is then established which would correspond to a predetermined pressure drop across the sample. The second phase, a gas, is then admitted to the sample at a low rate. After equilibrium has been established, the saturation is determined by weighing. Successively higher gas flow rates, and correspondingly lower rates of the first phase, are employed in succeeding steps to determine complete relative permeability–saturation relations. The single sample dynamic method is very simple to carry out but of questionable value for short samples.

In the *stationary liquid method* (Leas et al., 1950; Branson, 1951) boundary effects are avoided by holding the wetting phase stationary within the sample by capillary forces. Only the non-wetting phase is permitted to flow. This is accomplished by the use of a very low pressure gradient across the sample. The method is applicable only to the determination of relative permeability to the non-wetting phase, which must be gaseous. With this method, relative permeability–saturation curves for gas are obtained by starting with the sample completely saturated with wetting phase. The wetting phase saturation is then reduced by blotting the sample. When the wetting phase saturation becomes less than the equilibrium saturation, the wetting phase is passed upward through the sample under a small pressure drop. The relative permeability to the non-wetting phase at that point is measured, and the wetting phase saturation is determined by weighing. This process is repeated until the complete relation of the relative permeability of the non-wetting phase to saturation is obtained. This method is very simple to carry out but of rather questionable accuracy.

In the *gas drive method* the influence of the boundary effect is minimized by the use of a high rate of flow. The method differs from the stationary liquid method in that both phases flow simultaneously; it differs from the single sample dynamic method in that only the non-wetting phase, a gas, is admitted to the sample. The flow of both phases results from the displacement of portions of the wetting phase by the flowing gas: hence the name, "gas drive." This method was first suggested by Hassler et al. (1936). It is fairly good and very simple but occasionally gives faulty results.

Hassler (1944) also described another method for relative permeability measurements. The Hassler method has been modified by Gates and Lietz (1950), Brownscombe et al. (1950b) and Osoba et al. (1951). Its main features consist (after Osoba et al., 1951) in controlling the capillary pressure at both ends of the sample. This is accomplished by placing the sample between two discs permeable only to the wetting phase. This permits maintenance of a uniform saturation throughout the length of the sample, even at low rates of flow. The apparatus for measurements by the Hassler method is shown in figure 20. The semi-permeable discs at each end of the sample allow the wetting phase (a liquid) but not the non-wetting phase (a gas) to pass. They also permit the pressure drop in the wetting phase to be measured independently of that in the gas phase. In the measurement of relative permeability, the sample to be tested is saturated with wetting phase, and then is placed in the apparatus with a few layers of tissue paper at both ends to ensure good contact between the sample and semi-permeable discs. The wetting phase enters the sample through disc A. Disc A is isolated from the small disc B by the metal sleeve C. The pressure in the wetting phase at the inflow end is measured through the disc B. Gas enters the sample through radial grooves in the face of disc A. The difference between the pressure in the wetting phase and in the gas at the inflow end is the capillary pressure and is measured by the gauge p_c. The wetting phase and gas pass through the sample at rates such that the pressure drop in the two phases across the sample is the same. This is accomplished by adjusting valve V_G, which controls the rate of flow of gas through the sample, so that the pressure drop in the gas phase is equal to the pressure drop in the wetting phase. This is done so that the capillary pressure at the outflow end will be equal to the capillary pressure at the inflow end. The gas leaves the sample through radial grooves in disc D, and the pressure of the wetting phase at the outflow end is measured through disc E. After equilibrium is reached, the sample is removed and the saturation of the sample is determined by weighing. The Hassler method is exceedingly accurate but very complicated to carry out.

In the *Hafford technique* (after Osoba et al., 1951) the non-wetting fluid is fed directly into the sample and the wetting fluid is fed into the sample through a semi-permeable disc that allows only the wetting fluid to pass. The central portion of the semi-permeable disc is isolated from the remainder of the disc by a small metal sleeve. The central portion is used to measure the pressure in the wetting fluid at the input end of the sample. The pressure in the non-wetting fluid is measured through a standard pressure tap machined into the Lucite surrounding the sample. The pressure difference between the wetting and the

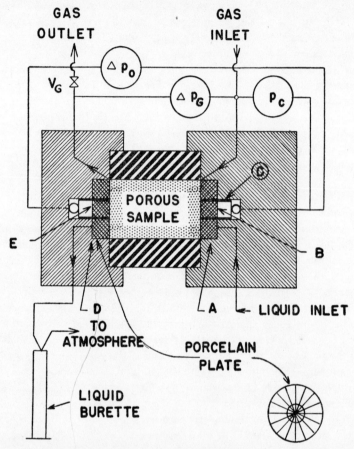

Figure 20. Hassler apparatus for relative permeability determination. (After Osoba *et al.*, 1951.)

non-wetting fluid is a measure of the capillary pressure of the sample at the inflow end. The Hafford technique gives excellent results and is fairly simple to carry out.

The *dispersed feed method* (after Richardson *et al.*, 1952) is similar to the Hafford and single sample dynamic methods. In this method the wetting fluid enters the test sample by first passing through a dispersing section. The dispersing section is made of porous material similar to the test sample, but it does not contain a device for measuring the pressure in the wetting fluid at the inflow end of the test sample as does the Hafford method. This porous material, which in some cases has been made from the same material as the test sample, serves to disperse the wetting fluid so that the wetting fluid enters the test sample more

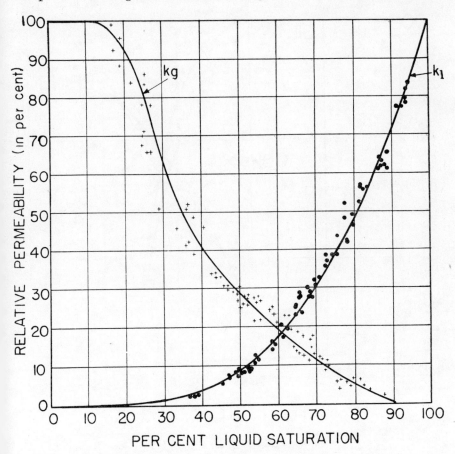

Figure 21. A typical example of relative permeability curves referring to a liquid (k_l) and a gas (k_g). (After Wyckoff and Botset, 1936.)

or less uniformly over the inflow face. Radial grooves are machined into the outflow face of the dispersing section. Gas is introduced to the test section through these radial grooves at the junction between the two sections. The dispersed feed method gives excellent results and is fairly simple to carry out.

Finally, the *seven methods* of relative permeability measurements are compared to each other in table V. For the convenience of the reader, the advantages and accuracy of each method are listed therein.

The various methods of measuring relative permeability have been applied on many occasions. Most such experiments are to find relative permeability curves for the flow of two or more phases in petroleum well cores. Such experiments and curves have been published (apart from the investigations already

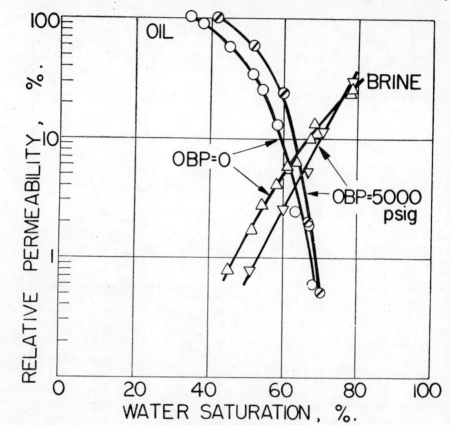

Figure 22. The effect of overburden pressure on relative permeability of an oil-brine system. (After Wilson, 1956.)

mentioned) for two-phase flow by Dunlap (1938), Krutter and Day (1943), Sen-Gupta (1943), and Henderson and Meldrum (1949); for three-phase flow by Evinger and Muskat (1942), and Holmgren and Morse (1951). Richards and Moore (1952) made similar experiments on soil samples, and Fatt (1953) studied the effect of overburden pressure on relative permeability.

According to these investigations, the relative permeability curves are invariably S-shaped curves. A typical example which has been determined by Wyckoff and Botset (1936) is supplied in figure 21. External parameters may, however, have an effect upon relative permeability. Thus it has been found that overburden pressure, for instance, not only changes the total permeability, but also the shape of the relative permeability curves. A beautiful set of experiments to demonstrate this fact has been undertaken by Wilson (1956); an example of his results is shown in figure 22.

As the relative permeability curves are always S-shaped curves, attempts have been made to give an analytical expression for them. Thus, Jones (1946, 1949) found that the following expression represents experimental data reasonably well in most cases:

$$k_n = 1 - 1.11 s_w \tag{8.2.3.1}$$

$$k_w = s_w^3 \tag{8.2.3.2}$$

where the subscript n stands for non-wetting phase, and the subscript w for wetting phase. The above relationship has been obtained from experiments involving oil as non-wetting phase, and water as wetting phase.

8.3. Solutions of Darcy's Law (Immiscible Fluids)

8.3.1. The Buckley-Leverett Case.
The system of equations (8.2.2.4) to (8.2.2.7) corresponding to Darcy's law, is extremely difficult to treat analytically, owing to its non-linear character. However, in some special cases solutions have been obtained.

An obvious solution has been given by Buckley and Leverett (1942). If gravity, capillarity and variations in density are neglected ("Buckley-Leverett assumptions") then it becomes evident that equations (8.2.2.5a/5c) become an integral of the motion.

We set

$$r(s) = \frac{|q_1|}{|q|} \tag{8.3.1.1}$$

such that $r(s)$ is the fractional flow of the 1-phase. Here, q is the total flow (equal to $q_1 + q_2$); the index "1" is omitted in the saturation as the equations are thought to pertain to the 1-phase only. It should be noted that $r(s)$ is indeed a function of saturation (and fluid viscosities) only, since one has in virtue of (8.2.2.4), if the gravity term is neglected and p_1 set equal to p_2:

$$r(s) = \frac{|q_1|}{|q_1 + q_2|} = \frac{k_1/\mu_1}{k_1/\mu_1 + k_2/\mu_2}. \tag{8.3.1.2}$$

As outlined in section 8.2.2, the relative permeabilities are functions of saturation only.

In view of the foregoing remarks, equation (8.2.2.5) can be written in the Buckley-Leverett case:

$$0 = P \frac{\partial s}{\partial t} + r'(s) \, (q \, \text{grad} \, s), \tag{8.3.1.3}$$

$$\text{div} \, q = 0 \tag{8.3.1.4}$$

where $r'(s) = dr/ds$.

In the one-dimensional case, equations (8.3.1.3/4) can be simplified to look as follows:

$$P\frac{\partial s}{\partial t} + r'(s)q_x\frac{\partial s}{\partial x} = 0, \qquad (8.3.1.5)$$

$$\frac{\partial q_x}{\partial x} = 0. \qquad (8.3.1.6)$$

Equation (8.3.1.6) gives at once:

$$q_x = q_x(t). \qquad (8.3.1.7)$$

Thus we have

$$P\frac{\partial s}{\partial t} + r'(s)q_x(t)\frac{\partial s}{\partial x} = 0. \qquad (8.3.1.8)$$

The last equation is a first-order differential equation and is amenable to treatment by the method of characteristics. A characteristic of a first-order partial differential equation is a line in the x-t plane along which the change of the unknown function (s) can be determined by a *total* differential equation. The change of s is:

$$ds/dt = (\partial s/\partial x)dx/dt + \partial s/\partial t. \qquad (8.3.1.9)$$

If the ratio between dx and dt is chosen in such a manner that the right-hand side of equation (8.3.1.9) becomes equal to the left-hand side of equation (8.3.1.8), then the change ds is determined by a total differential equation. Obviously, dx/dt has to be chosen equal to the ratio of the coefficients in equation (8.3.1.8) associated with $\partial s/\partial x$ and $\partial s/\partial t$ respectively. We thus find that along the "characteristic lines" determined by

$$dx/dt = r'(s)q_x(t)/P \qquad (8.3.1.10)$$

the change of s is given by the right-hand side of equation (8.3.1.8), viz.

$$ds/dt = 0. \qquad (8.3.1.11)$$

Thus we have succeeded in replacing the partial differential equation (8.3.1.8) by two total differential equations (8.3.1.10) and (8.3.1.11). The last two equations are the equations of Buckley and Leverett. These authors arrived at (8.3.1.10) and (8.3.1.11) from (8.3.1.8) by "dividing out" *partial* differentials, which is not always permissible as it may give a wrong result. In the present case, however, the rigorous derivation of (8.3.1.10) and (8.3.1.11) shows that the Buckley-Leverett equations are correct. It is quite immaterial whether in the partial differential equation (in our case equation (8.3.1.8)) the right-hand side is zero or not; the method of characteristics works in either case. In the present case, where the right-hand side of (8.3.1.8) *is* zero, we have the additional convenience that the right-hand side of (8.3.1.11) is also zero and the characteristics (equation (8.3.1.10)) are therefore lines in the x-t plane along which there is a constant saturation.

It is thus seen that any saturation proceeds with the speed given by the tangent to the $r(s)$-curve at that saturation (up to a factor which is constant for any given time, and constant *all* the time if the total flow is kept constant).

Brinkman (1948) has shown that equations (8.3.1.10) and (8.3.1.11) have solutions which take the form of a shock: a finite discontinuity of saturation is proceeding at a speed c. Although it is very difficult to prove that such a shock will always develop when the displacing fluid is injected at $x = 0$ at saturation 1 (thus one has saturation zero for the 1-phase to which our "s" refers), it is very easy to see that such solutions *do* exist and to give their explicit form.

Thus, let it be assumed that there is such a shock. To assist visualization, let it be assumed that the displacement takes place from left to right. To the right of the shock there is $s = 1$, to the left of the shock, there is some saturation s; to the right of the shock, $r = 1$, to the left some r (r and s behind the shock are to be determined). Thus, r and s refer to the *displaced* fluid.

We can formulate the continuity condition across the shock as follows:

$$Pc(1 - s) = q_x(1 - r); \qquad (8.3.1.12a)$$

this is termed "shock-condition" (Hugoniot equation). We obtain from it at once the speed of the shock

$$c = \frac{q_x}{P} \frac{1-r}{1-s}. \qquad (8.3.1.12b)$$

Equation (8.3.1.10) gives the distance travelled by the various saturations of the displaced fluid in any given time. The distance travelled is proportional to the $r'(s)$ function, the "constant of proportionality" being equal to $\int q_x(t)dt/P$. Thus, a plot of s versus r', such as the one in figure 23 for a concrete case (see Welge, 1952) also gives the saturation of the displaced fluid as a function of distance. It is only necessary to multiply the $r'(s)$ scale by the value $\int q_x(t)dt/P$ corresponding to the time at which the saturation distribution is desired. Thus, the curve describing saturation as a function of distance remains always similar except for horizontal "stretching."

In the event that the velocity and the saturation at the farthest point (i.e., the "breakthrough saturation") is desired, Welge (1952) has shown that one can draw a tangent to the $r(s)$ curve from the point $r = 1$, $s = 1$, as shown in figure 23, for it is quite clear that, if the saturation immediately behind the shock travels more slowly than the shock, it will take only a very short time until the size of the shock is decreased to a saturation which travels at least as fast as the shock. Once the saturation immediately behind the shock travels with the same speed as the shock, it will remain that way: the same saturation will always stay just behind the shock. This saturation, therefore, is the breakthrough saturation. The speed of any saturation is given by equation (8.3.1.10). The speed of the

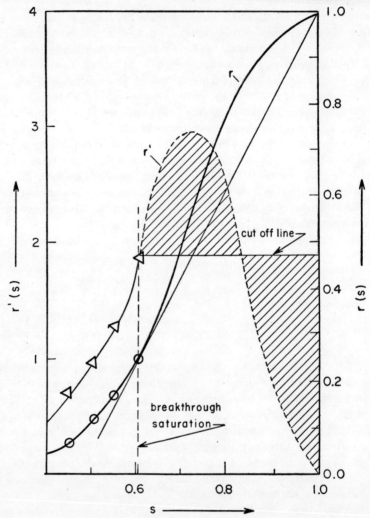

Figure 23. Fractional flow and its derivative in the Buckley-Leverett case. (After Welge, 1952.)

breakthrough saturation is c, as given by equation (8.3.1.12b). The condition for the breakthrough saturation is therefore obtained by equating the right hand sides of equation (8.3.1.10) and (8.3.1.12b) which proves the validity of the tangent construction outlined above.

Instead of using the tangent construction of Welge, the breakthrough saturation can also be obtained by cutting off the $r'(s)$ curve (as indicated in figure 23) by making the two shaded areas of equal size (Brinkman, 1948). Because of elementary properties of integrals, this construction is obviously equivalent to the tangent construction.

The problem of finding the saturations is now completely solved. If a displacing fluid be injected with (seepage) velocity q_x into a linear system, a bank forms where the saturation drops from 1 (of the displaced fluid) to the breakthrough saturation. The bank stays the same at all times and travels with speed c given by equation (8.3.1.12b). The saturations behind the bank travel with their characteristic speeds given by equation (8.3.1.10) and their advance can be found for any desired instant.

The general aspects of the Buckley-Leverett case have been described on several occasions, for example by Muskat (1949b) and Calhoun (1953).

The fundamental solution obtained by Buckley and Leverett has been extended for more complicated occasions. Tarner (1944) applied it to gas flow in oil fields and calculated a few special cases. Nielsen (1949) used an analytical (corresponding to equation 8.2.3.1/2) relative permeability formula and obtained a complete analytical solution of the Buckley-Leverett case. Muskat (1950) investigated the problem of stratified layers of different permeability. Kern (1952) analysed the displacement mechanism in multiwell systems.

Finally, the influence and possible carrying of the gravity term have been investigated by Lewis (1944), Brinkman (1948), and Terwilliger *et al.* (1951). The fundamental equations are again obtained by writing down the two Darcy equations and the two continuity equations for the two phases.

8.3.2. Other Analytical Solutions; Elementary Treatment. The solution of the linear flood of Buckley and Leverett, without accounting for capillarity, gravity, and density variations is but a very particular solution of the general problem stated by equation (8.2.2.4/7). It cannot even be generalized to two dimensions. In two dimensions, the eliminant corresponding to equation (8.3.1.8) contains three independent variables; the characteristics therefore become surfaces. It can be seen that, rather than proceed in this manner, a hodograph transformation is a better way to treat the problem. However, the real difficulty is to formulate the shock condition correctly. The latter now becomes a floating boundary condition, depending on the position of the shock *surface* (not point only). It is difficult to see how this can be handled adequately; in all probability recourse will have to be taken to methods of functional analysis. At any rate, at the present time an easy solution of a truly two-dimensional shock problem cannot be hoped for.

Therefore, a series of "solutions" have been based upon very stringent simplifications. Thus, the discussion of single phase flow in porous media may be extended to a consideration of multiple phase flow if simplifying approximations are made. If, for example, a fluid is displaced in a porous medium by another in such a manner that the "input" and the "output" are conducted steadily and if, furthermore, the permeability-viscosity ratio is the same for both fluids, and if no mixing occurs at the interface, then one can argue that the flow of the two fluids would occur in the same manner as if they were one homogeneous fluid. The position of the interface at an arbitrary time t could be ascertained from a consideration of its position at the time t_0 and a study of the motion of each point of the interface during the time from t_0 to t.

It is not even necessary to assume that the permeability-viscosity ratio for the two fluids is identical, and it is then still possible to apply elementary steady-state flow considerations to multiple phase flow, provided no mixing occurs at the interface and provided the permeability-viscosity ratios are constant. Muskat (1934) has given an analytical formulation to this problem.

Let the interface be represented at the time t by the equation

$$F(x, y, z, t) = 0. \tag{8.3.2.1}$$

Then at the time $t + \delta t$ the new interface will be given by

$$F(x + q_x \delta t,\quad y + q_y \delta t,\quad z + q_z \delta t,\quad t + \delta t) = 0 = F(x, y, z, t) \tag{8.3.2.2}$$

where q_i are the filter velocity components on the interface. We have at once:

$$\partial F/\partial t + \mathbf{q} \operatorname{grad} F = 0. \tag{8.3.2.3}$$

Applying now Darcy's law, we find

$$\frac{\partial F}{\partial t} - \frac{k}{\mu} \operatorname{grad} \Phi \operatorname{grad} F = 0 \quad \text{with} \quad \Phi = p + \rho g z. \tag{8.3.2.4}$$

The problem may now be formulated as follows (Muskat, 1934). Determine the potential functions Φ, between a surface S_w and $F(x, y, z, t) = 0$, and Φ_2 between a surface S_e and $F(x, y, z, t) = 0$, such that

$$\left. \begin{array}{r} \left. \begin{array}{l} \Phi_1 = \Phi_w \text{ on } S_w \\ \Phi_2 = \Phi_e \text{ on } S_e \end{array} \right\} \text{ boundary conditions} \\[1em] \left. \begin{array}{l} \Phi_1 = \Phi_2 \\ \dfrac{k_1}{\mu_1} \dfrac{\partial \Phi_1}{\partial n} = \dfrac{k_2}{\mu_2} \dfrac{\partial \Phi_2}{\partial n} \end{array} \right\} \text{ on } F(x, y, z, t) = 0 \\[1em] \dfrac{\partial F}{\partial t} - \left(\dfrac{k}{\mu} \right)_{1,2} \operatorname{grad} \Phi_{1,2} \operatorname{grad} F = 0 \end{array} \right\}. \tag{8.3.2.5}$$

Here, n is the normal to the interface. The fact that Φ has been assumed to be a potential function implies, of course, that steady state conditions (i.e., incompressibility) are maintained in each of the two flowing phases.

Even the very simplified problem outlined above is difficult to treat analytically and only a few solutions have been obtained. Muskat (1934) solved, for example, the case of radial encroachment in two dimensions upon this basis. The shape of the interface is known in advance in this case; it is always a circle. Let its position be denoted by r_0, and, furthermore, let the boundaries S_w and S

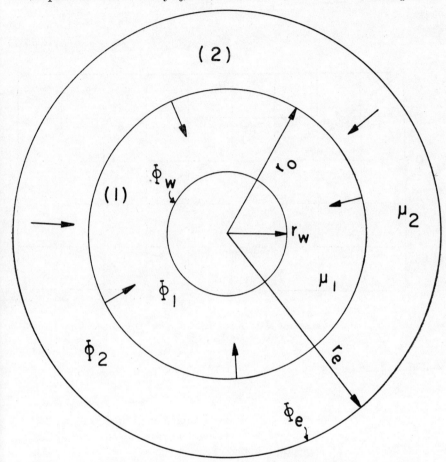

Figure 24. Geometrical layout in Muskat's approximate solution of a radial displacement problem. (After Muskat, 1934.)

also be circles of radius r_w and r_e ($r_w < r_0 < r_e$), respectively (see figure 24). If the potentials are held constant upon the boundaries ($= \Phi_w$ and Φ_e, respectively), then the equation for the interface can be integrated readily and we obtain:

$$\frac{r_0^2}{r_e^2}\left\{\log\text{nat}\frac{r_e^2}{r_w^2} - (1-\varepsilon) + (1-\varepsilon)\log\text{nat}\frac{r_0^2}{r_e^2}\right\} = -4\frac{k_1}{\mu_1}t\frac{\Phi_e - \Phi_w}{r_e^2}$$

$$+ \log\text{nat}\frac{r_e^2}{r_w^2} - (1-\varepsilon), \quad (8.3.2.6)$$

with

$$\varepsilon = \frac{k_1 \mu_2}{\mu_1 k_2}. \qquad (8.3.2.7)$$

Setting

$$4 \frac{k_1}{\mu_1} (\Phi_e - \Phi_w) \cdot \frac{1}{r_e^2} = 1, \qquad (8.3.2.8)$$

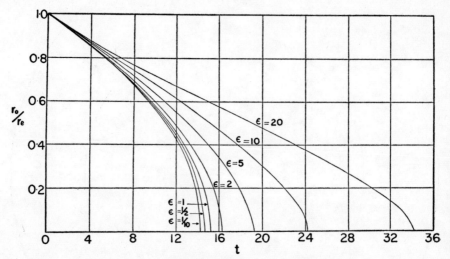

Figure 25. Muskat's approximate solution of a radial displacement problem. (After Muskat, 1934.)

and adding the numerical assumption that

$$r_e/r_w = 2000, \qquad (8.3.2.9)$$

curves representing equation (8.3.2.6) are obtained as plotted in figure 25.

More complicated geometrical patterns can be treated more easily by setting, in addition to the above simplifying assumptions,

$$\frac{k_1}{\mu_1} = \frac{k_2}{\mu_2} \qquad (8.3.2.10)$$

which implies that there is, in essence, a single-phase flow problem.

Muskat in his book (1937) discusses a variety of solutions for heterogeneous flow based upon this basis. Many solutions of the two-phase problem based upon this basis were also given in a later book by Muskat (1949). The original papers on this subject commence as early as 1933 with a report by Wyckoff et al. (1933) on the mechanics of porous flow applied to water flooding networks. Similar studies have been made by Muskat and Wyckoff (1934, 1935) and by Wyckoff and

Botset (1934). Hubbert (1940) reviewed some cases in his article on groundwater motion, and Braginskaya (1942) studied displacement upon this elementary basis in anisotropic media. Furthermore, similar problems have been treated by Polubarinova-Kochina (1943) and by Yuster and Calhoun (1944). Yuster (1946) gave nomographs for the prediction of water flooding intakes in oil fields, Kazarnovskaya (1947) investigated the movement of the water-oil interface and water encroachment under a hydrostatic head, and Muskat (1947) studied the performance of bottom water-drive oil pools. Other calculations of oil displacement by water in oil fields upon this elementary basis have been reported by Muskat (1948), Ryder (1949c), Stiles (1949), Suder and Calhoun (1949), Adamson and Brown (1953), and Dietz (1953).

The heterogeneous flow theory based upon the above-mentioned simplifying assumptions is actually only a crude approximation to the physical facts. Although it is possible that the factor k/μ is identical for both fluids if they are considered singly, the condition of no mixing at the interface is never satisfied. In the region where mixing occurs, the permeability of the medium to the mixture of the two fluids is much lower than that to the single fluids so that the elementary theory outlined above is, to this extent, incorrect. Nevertheless, it seems to be capable of giving qualitative results for industrial and engineering applications such as the sweep efficiency of certain types of arrays of injection and withdrawal wells in an oilfield. However, the details of the calculations are not very relevant as far as the discussion of the physics of displacement is concerned. The reader is therefore referred to the original papers if more detailed information is desired.

8.3.3. Imbibition. A certain solution of the basic equations describes a peculiar phenomenon which is known as imbibition of a wetting fluid into a porous medium. This phenomenon has been discussed by Brownscombe and Dyes (1952) and by Enright (1954).

Assume* that a cylindrical piece (linear co-ordinate x) of porous matrix is completely surrounded by an impermeable surface except for one end of the cylinder which is designated as the imbibition face. If the matrix is completely filled with some fluid, then it is observed that the presence of a more wetting fluid against the imbibition face gives rise to a flow of wetting fluid into the matrix and a counterflow of original fluid from the matrix.

If it is assumed that both the flow of wetting fluid (w) and the counterflow of original fluid (n) are governed by Darcy's law, we may write

$$q_w(x, t) = (k_w(x, t)/\mu_w)k\partial p_w(x, t)/\partial x \qquad (8.3.3.1)$$

$$q_n(x, t) = (k_n(x, t)/\mu_n)k\partial p_n(x, t)/\partial x. \qquad (8.3.3.2)$$

---

* The following discussion of the imbibition phenomenon is taken from unpublished work by J. G. Richardson and J. W. Graham of the Humble Oil Co. in Houston, Texas. The writer is indebted to the Humble Oil Company for giving him access to this work.

Also, the capillary pressure is related to the non-wetting and wetting fluid pressures by

$$p_c = p_w - p_n. \tag{8.3.3.3}$$

Combining (8.3.3.1), (8.3.3.2), and (8.3.3.3), noting that $q_n(x, t) = - q_w(x, t)$, we obtain:

$$q_n(0, t) = - k \left[\frac{k_w(0, t)k_n(0, t)}{\mu_w k_n(0, t) + \mu_n k_w(0, t)}\right]\left[\frac{\partial p_c}{\partial x}\right]_{x=0}. \tag{8.3.3.4}$$

Equation (8.3.3.4) gives the outflow rate of non-wetting fluid from the imbibition face as a function of time.

According to Leverett (cf. equation 3.4.1.9), a definite relation exists between the dimensionless quantity $(p_c/\gamma)(k/P)^{\frac{1}{2}}$ and s_w, the wetting fluid saturation. We may express this relationship by writing (cf. equation 3.4.1.9)

$$J(s_w) = \frac{p_c}{\gamma}\sqrt{\frac{k}{P}}. \tag{8.3.3.5}$$

Equation (8.3.3.5), solved for p_c, yields

$$p_c = J(s_w)\gamma\sqrt{\frac{P}{k}}. \tag{8.3.3.6}$$

Combining equations (8.3.3.4) and (8.3.3.6) gives

$$q_n(0, t) = - \sqrt{kP} \left(\frac{k_w k_n}{\mu_w k_n + \mu_n k_w}\right)_{x=0} \left(\frac{\partial \gamma}{\partial x} J(s_w) + \gamma \frac{dJ}{ds_w}\frac{\partial s_w}{\partial x}\right)_{x=0} = 0. \tag{8.3.3.7}$$

The last equation is the expression of Darcy's law for the case under consideration. The continuity condition for the same case has also to be formulated. We could proceed from the basic equations as listed in section 8.2.2, but it is just as easy to write down the balance for a slice of thickness dx of the cylinder:

$$q_w(x, t)dt - q_w(x + dx, t)dt = P[s_w(x, t + dt)dx - s_w(x, t)dx]. \tag{8.3.3.8}$$

Dividing both sides of (8.3.3.8) by $(dxdt)$ and passing to the limit, we obtain

$$P(\partial s_w/\partial t)_x = - (\partial q_w/\partial x)_t, \tag{8.3.3.9}$$

which is the equation of continuity. This equation also implies that

$$\left[\frac{\partial s_w}{\partial x}\right]_t = - \frac{(\partial q_n/\partial x)_t}{P(\partial x/\partial t)_{s_w}}. \tag{8.3.3.10}$$

Combining equations (8.3.3.7) and (8.3.3.10), we obtain

$$q_n(0, t) = \sqrt{k} \left(\frac{k_w k_n}{\mu_w k_n + \mu_n k_w}\right)_{x=0} \left(\sqrt{P}\frac{\partial \gamma}{\partial x} J(s_w) - \frac{\gamma}{\sqrt{P}}\frac{\partial J}{\partial s_w}\frac{(\partial q_n/\partial x)_t}{(\partial x/\partial t)_{s_w}}\right)_{x=0}. \tag{8.3.3.11}$$

Equation (8.3.3.11) permits a comparison with data that can be obtained in the laboratory for the amount of original fluid produced (Q) by imbibition as a function of time, namely:

$$Q(t) = \int_0^t q_n(0, t)dt. \qquad (8.3.3.12)$$

Although this integration is difficult to attempt, several important features of the imbibition phenomenon are brought to light:

(1) the outflow rate of non-wetting fluid, $q_n(0, t)$ varies directly with $k^{\frac{1}{2}}$ all other factors remaining unaltered;
(2) the term in the first brackets can be written as

$$\frac{k_w k_n}{\mu_w k_n + \mu_n k_w} = \frac{k_n}{\mu_w(k_n/k_w) + \mu_n}. \qquad (8.3.3.13)$$

Since the region of matrix in the vicinity of the imbibition face soon acquires high wetting fluid saturation, it is a good approximation to write:

$$\frac{k_w k_n}{\mu_w k_n + \mu_n k_w} = \frac{k_n}{\mu_n}. \qquad (8.3.3.14)$$

Therefore, the outflow rate $q_n(0, t)$ varies nearly as $1/\mu_n$.

8.3.4. Solution Subject to Constraints. In case there is an energy relationship between two flowing phases in a porous medium, the methods outlined above have to be modified. In general, such a relationship acts as a constraint in the solution.

Such a condition might occur if the liquid (index w) is displaced by its vapour (index G). The solution of the problem may be attempted for a linear system (linear co-ordinate x) as follows (suggested by J. von Neumann).

The continuity equation for mass yields

$$0 = P \frac{\partial}{\partial t} \{s_w \rho_w + (1 - s_w)\rho_G\} + \frac{\partial}{\partial x} \{r_w \rho_w + (1 - r_w)\rho_G\} q. \qquad (8.3.4.1)$$

A similar continuity equation must hold for the heat flow:

$$\frac{\partial}{\partial t} \{P(\zeta_w s + \zeta_G(1 - s) + \zeta_R\} + HL = -\frac{\partial}{\partial x} \{q(\zeta_w r + \zeta_G(1 - r))\} \qquad (8.3.4.2)$$

Here, ζ is the heat per unit mass for liquid, vapour, and porous matrix (corrected for porosity) denoted by the index w, G, R respectively, as a function of temperature; HL is a heat-loss term. All the other symbols have the same meaning as before. If an index is omitted, it is understood that the corresponding term refers to *liquid*.

The problem can only be solved sensibly if there exists a front for the advance of the vapour (a bank in the usual sense). In that case the temperature in front of the bank will be $\mathfrak{T}_0$, i.e., the original temperature of the system; behind the bank it will be $\mathfrak{T}$, i.e., that of the vapour. Under these conditions, the heat-loss term is

$$HL = \frac{2C(\mathfrak{T} - \mathfrak{T}_0)}{h\sqrt{\pi}\sqrt{t-\tau}} \qquad (8.3.4.3)$$

where C is some heat constant and h the width of the system. If the velocity of the bank is denoted by c, then we have

$$\tau = x/c \qquad (8.3.4.4)$$

if c is kept constant. (The "injection" of vapour can always be arranged in such a fashion.) If c is not kept constant, we have to integrate in order to obtain τ.

The densities of all the phases involved are assumed to be constant. The heats per unit mass (ζ) refer now to the heat necessary to raise the temperature from $\mathfrak{T}_0$ to $\mathfrak{T}$ in the respective phase; the heat for the gas includes the heat of evaporation. These heats (ζ) are assumed as constant.

Under these assumptions, we can write the two continuity conditions, valid anywhere *behind* the shock front (i.e., at temperature $\mathfrak{T}$) as follows:

$$P\frac{\partial}{\partial t}[\rho_w s + \rho_G(1-s)] = -\frac{\partial}{\partial x}[q\{\rho_w r + \rho_G(1-r)\}]. \qquad (8.3.4.5)$$

$$P\frac{\partial}{\partial t}\{\zeta_w s + \zeta_G(1-s) + \zeta_R\} + \frac{2C(\mathfrak{T} - \mathfrak{T}_0)}{h\sqrt{\pi}\sqrt{t-\dfrac{x}{c}}} = -\frac{\partial}{\partial x}\{q[\zeta_w r + \zeta_G(1-r)]\}. \qquad (8.3.4.6)$$

Following the method of Buckley and Leverett, an eliminant of those two equations can be constructed. We shall use the following abbreviations:

$$\alpha = \frac{\rho_G}{\rho_w - \rho_G}; \quad \beta = \frac{\zeta_G}{\zeta_w - \zeta_G}, \qquad (8.3.4.7)$$

$$\sigma = \frac{2C(\mathfrak{T} - \mathfrak{T}_0)}{h\sqrt{\pi}(\zeta_w - \zeta_G)}. \qquad (8.3.4.8)$$

It is, then, obvious that the continuity equations can be written in the following manner

$$P\frac{\partial s}{\partial t} = -\frac{\partial}{\partial x}[(r + \alpha)q], \qquad (8.3.4.9)$$

$$P\frac{\partial s}{\partial t} = -\frac{\partial}{\partial x}[(r + \beta)q] - \frac{\sigma}{\sqrt{t - \dfrac{x}{c}}}. \qquad (8.3.4.10)$$

§8.3   Solutions of Darcy's Law

This allows the construction of the following eliminant

$$\frac{\partial q}{\partial x} = \frac{\sigma}{(\alpha - \beta)\sqrt{t - \frac{x}{c}}} \tag{8.3.4.11}$$

so that one obtains, with the correct constant of integration,

$$q = q(t) + \frac{2c\sigma}{\alpha - \beta}\left(\sqrt{t} - \sqrt{t - \frac{x}{c}}\right). \tag{8.3.4.12}$$

This can be inserted into the first continuity equation (8.3.4.9) to yield

$$P\frac{\partial s}{\partial t} = -qr'\frac{\partial s}{\partial x} - (r+\alpha)\frac{\sigma}{\alpha-\beta}\frac{1}{\sqrt{t - \frac{x}{c}}} \tag{8.3.4.13}$$

with

$$r' = dr/ds. \tag{8.3.4.14}$$

Using the method of characteristics, we obtain therefore

$$\left.\frac{ds}{dt}\right|_{\left[\frac{dx}{dt} = \frac{1}{P}qr'\right]} = -(r+\alpha)\frac{\sigma}{\alpha-\beta}\frac{1}{\sqrt{t - \frac{x}{c}}}. \tag{8.3.4.15}$$

An analysis of the actual numerical values of α and β shows that for all physically sensible cases the following conditions hold:

$$\alpha > 0, \quad \beta > 0, \quad \alpha - \beta < 0 \tag{8.3.4.16}$$

so that the right-hand side of the characteristic equation is positive.

The equation for the characteristics gives the complete solution of the problem, except for the relationship between c and q. The latter is deduced from the "shock condition," i.e., the continuity condition around the bank.

Thus, the shock condition is obtained by formulating, for example, the heat-balance equation across the shock front. This procedure yields

$$cP(\zeta_w s + \zeta_G(1-s) + \zeta_R) = q(\zeta_w r + \zeta_G(1-r)). \tag{8.3.4.17}$$

After some algebraic transforming, this yields

$$c = \frac{q}{P}\frac{r+\beta}{s+\beta'}, \tag{8.3.4.18}$$

with

$$\beta' = \beta + \frac{\zeta_R}{\zeta_w - \zeta_G}. \tag{8.3.4.19}$$

All this works only if c is kept constant, which is a working assumption. Otherwise, we have to introduce integrals for the calculation of τ.

The problem (with constant c) is now amenable to a graphico-numerical solution, The r versus s curve is experimentally known; it is an S-shaped curve, starting out from the origin. The speed of the front (up to the factor q/P) is given by the secant from the point $(-\beta, -\beta')$ to a point "X" on the curve, representing the instantaneous saturation at the front. If the tangent to the curve at the point X is steeper than this secant, then the saturations around the bank can be determined at the time $t + dt$ (where the time t corresponds to the point X). This is clear because the characteristics travel with the speed dr/ds (up to the factor q/P); thus, the "information" travels faster than the bank.

From the point $(-\beta, -\beta')$ it is possible to draw two tangents to the curve (the latter being S-shaped); call the point of contact of the lower tangent "I", the point of contact of the upper tangent "II." We run into difficulty if X is not between the points I and II on the r versus s curve. It can be shown that if X is between 0 and I, it will go into I in an infinitesimal time; once at I it continues to ascend the curve. It is not at all clear, however, whether it will ever attain II; if it does, it would further ascend the curve and the problem would be undetermined. It is likely that in such a case a second bank would form.

8.4. Scaling and Experimental Investigations into Flow Patterns of Immiscible Fluids in Porous Media

8.4.1. Dynamical Similarity in Porous Media.
The problem of dynamical similarity in porous media is important if one wants to scale flow phenomena. This problem has been discussed by Rapoport and Leas (1953) who solved it for a special case of two-fluid flow of immiscible fluids in porous media, *viz.*, for the case of linear flow, neglecting density variations in either fluid as well as gravity. The theory was extended later.

Following Rapoport and Leas (1953), it is easy to see that, from the fundamental equations of multiple phase flow in porous media (8.2.2.4/7), the following eliminant can be formed, provided the two fluids are incompressible and gravity is neglected:

$$P \frac{\partial s_1}{\partial t} + r'(s_1) q \operatorname{grad} s_1 - \frac{k}{\mu_2} \operatorname{div} (k_2 r(s_1) p_c' \operatorname{grad} s_1) = 0 \qquad (8.4.1.1)$$

where r is the fractional flow of the one-phase (as defined in 8.3.1.2) and the other symbols have the same meaning as in section 8.2.2. If an index is omitted, it is understood that the corresponding term refers to the one-phase.

By means of the above eliminant, scaling problems can be studied. The eliminant equation can be brought into dimensionless form if the following substitutions of variables are made:

$$X = x/L \text{ (etc. for } y, z); \qquad (8.4.1.2)$$

$$T = t|q|/LP \qquad (8.4.1.3)$$

where L is some characteristic macroscopic length of the flow system. This yields

$$\frac{\partial s_1}{\partial T} + r'\boldsymbol{n}\,\text{grad}_X\,s_1 - \frac{k\mu_1}{\mu_2}\frac{1}{L|q|\mu_1}\,\text{div}_X\,[k_2 r p_c'\,\text{grad}_X\,s] = 0 \qquad (8.4.1.4)$$

where $\boldsymbol{n} = \boldsymbol{q}/|q|$. This equation being in dimensionless form, it is seen that dynamical similarity is obtained if the "scaling" coefficient $\mathfrak{Z}$

$$\mathfrak{Z} = L|q|\mu_1 \qquad (8.4.1.5)$$

as well as the viscosity ratio μ_2/μ_1 have identical values in two flow systems. Naturally, the two systems must also be geometrically similar and must consist of identical porous media.

For small values of the scaling coefficient $\mathfrak{Z}$, examination of the dimensionless eliminant equation (8.4.1.4) shows that the second order term, containing the capillary pressure, is relatively large with respect to the two first order terms. Under such circumstances, the flow behaviour is influenced by capillary pressure forces and is markedly dependent upon rate, length of flow system, and fluid viscosities. As the value of the scaling coefficient increases, it is, however, seen that the relative importance of the second order term diminishes and that it will eventually become negligible. The eliminant equation then reduces to a simplified form, in which only first order terms are retained:

$$\frac{\partial s_1}{\partial T} + r'\boldsymbol{n}\,\text{grad}_X\,s_1 = 0. \qquad (8.4.1.6)$$

Expression (8.4.1.6) is independent of L, q, and the viscosities, provided the viscosity *ratio* remains unchanged. Accordingly, displacement experiments ("floods") may be defined as "stabilized," if they are independent of the rate of injection, the length of the flooded system, and the fluid viscosities. This is the case if the scaling coefficient $\mathfrak{Z}$ is large.

One can extend the above by starting from the general flow equations. The following eliminant can be formed from equations (8.2.2.4/7) (Rapoport, 1954):

$$P\frac{\partial s_1}{\partial t} + r'(s_1)\boldsymbol{q}\,\text{grad}\,s_1 - \frac{k}{\mu_2}\,\text{div}\,(k_2 r p_c'\,\text{grad}\,s_1) - \frac{k}{\mu_2}\,g(k_2 r)'\Delta\rho\,\frac{\partial s_1}{\partial z} = 0$$

$$(8.4.1.7)$$

where $\Delta\rho$ is the density difference between the two flowing fluids (defined as $\rho_1 - \rho_2$), the latter being again assumed as incompressible. However, the gravity term is now not being neglected.

As before, the eliminant equation (8.4.1.7) makes possible a study of scaling problems. Denote the "original" by capital letters, the "model" by lower case ones. We can then introduce the following scaling factors:

$$X/x = Y/y = Z/z = A \text{ (linear dimensions)} \tag{8.4.1.8}$$

$$U/u = V/v = W/w = C \text{ (filter velocities)} \tag{8.4.1.9}$$

$$N/n = B \text{ (volume flow rates)} \tag{8.4.1.10}$$

$$K/k = D \text{ (permeabilities)} \tag{8.4.1.11}$$

$$P_c/p_c = E \text{ (capillary pressures)} \tag{8.4.1.12}$$

$$\Delta P/\Delta\rho = F \text{ (density differences)} \tag{8.4.1.13}$$

$$T/t = G \text{ (times)} \tag{8.4.1.14}$$

$$M/\mu = H \text{ (viscosities)} \tag{8.4.1.15}$$

$$P/p = R \text{ (porosities)} \tag{9.4.1.16}$$

If, now, the eliminant equation (8.4.1.7) is used to determine scaling factors, we obtain at once:

$$ADE = BH = A^2DF \tag{8.4.1.17}$$

or, preferably written,

$$E = AF \tag{8.4.1.18}$$

$$BH = A^2DF \tag{8.4.1.19}$$

It may be noted that the scaling of times and velocities is not arbitrary; we have

$$C = B/A^2 \tag{8.4.1.20}$$

$$G = A^3R/B. \tag{8.4.1.21}$$

From a macroscopic viewpoint, the conditions expressed by equations (8.4.1.18/19) are sufficient to achieve proper scaling between an "original" and a corresponding "model." However, it was simply assumed that the relative permeability curves of wetting and non-wetting fluid (not the total permeabilities) are identical in model and original. It seems desirable to investigate the conditions that render this the case, considering the microscopic aspects of flow through porous media. Thus, we must also discover what similarity conditions apply in a microscopic sense to properties such as pore geometry, capillary pressure curves, etc.

Rapoport (1954) has given a discussion of this problem, but based upon the Kozeny equation (cf. section 6.3.2). As stated earlier, the Kozeny equation makes use of very definite assumptions about the constitution of porous media, which we have maintained earlier are somewhat doubtful. The same results, however, can be obtained by much less stringent ideas about the constitution of porous media as shown by Miller and Miller (1956).

Thus, assuming that one has two porous media which are geometrically similar, i.e., identical up to a scaling factor λ (this scaling factor being applied microscopically to the pore geometry), the dynamics within are characterized by two kinds of laws: (i) the surface tension law, and (ii) the (microscopically applied) Navier-Stokes equation.

The surface tension equation (3.4.1.3),

$$\frac{4 \cos \theta}{\delta_c} = \frac{p_c}{\gamma} \tag{8.4.1.22}$$

(see section 3.4.1 for definition of symbols), indicates by its form that if the surface tension γ is doubled and the variable p_c is everywhere doubled, the interface shapes (i.e., their radii of curvature δ_c) consistent with (8.4.1.22) will be unchanged. The porosity does not enter into these considerations. Thus, in order to make interfaces between two phases in geometrically similar porous media, similar also, we have to choose, assuming identical contact angles θ:

$$\frac{1}{\lambda} = \frac{P_c/p_c}{\Gamma/\gamma} = \frac{E}{S} \tag{8.4.1.23}$$

or

$$S = E\lambda \tag{8.4.1.24}$$

where the various scaling factors are defined as follows:

$$\Delta_c/\delta_c = \lambda \text{ (microscopic lengths, radii of curvature)}, \tag{8.4.1.25}$$

$$\Gamma/\gamma = S \text{ (surface tensions)}. \tag{8.4.1.26}$$

Furthermore, the Navier-Stokes equation (2.2.2.1) yields that, for a given liquid geometry, the resistance to flow will vary with the viscosity μ and vary inversely with δ_c^2. Thus, averaging over the sample, the permeability to a phase will vary directly with δ_c^2, i.e.:

$$kk_1 = \text{const. } \delta_c^2. \tag{8.4.1.27}$$

(In connection with this statement cf. also section 6.5.3 on statistical theories of permeability.) Thus:

$$D\kappa_1 = \lambda^2 \tag{8.4.1.28}$$

where the scaling factor κ_1 is defined as follows:

$$K_1/k_1 = \kappa_1 \text{ (relative permeabilities)}. \tag{8.4.1.29}$$

Since the macroscopic eliminant equation requires that κ_1 equal 1, we obtain as microscopic similarity conditions:

$$D = \lambda^2 = S^2/E^2. \qquad (8.4.1.30)$$

This relation was also found, by Rapoport (1954), thus demonstrating that in its deduction the particular ideas introduced by the application of the Kozeny equation, would actually have been justified.

8.4.2. Experimental Displacement Studies. Owing to the analytical difficulties of solving the multiple phase flow Darcy equations, attention has been directed towards the possibility of experiments at a very early stage. Unfortunately, the idea of scaling outlined in 8.4.1 has not been applied to such experiments until very recently, although the general principles of dynamical similarity have been well known in physics for about a century.

In fact, the only experiments on two-fluid flow that were performed were undertaken by the oil industry in order to determine the "recovery" of oil under various conditions of "gas-" or "water-flooding." The reports on the influence of the various variables on flooding experiments yielded, however, very erratic results—as was to be expected, since scaling had been neglected. The results obtained in the models, equally, are often not at all significant in relation to the condition in the "field" which they are supposed to represent. There is one saving factor, however. This is that the "scaling coefficient" $Lq\mu$ is often (inadvertently) very large in both the "model" and in the "field" so that the corresponding experiments are very often dynamically similar, i.e., the flood experiments are "stabilized" in the sense described in (8.4.1).

Thus, most experimental investigations are nothing but straightforward "floods" (using water or gas) on petroleum well cores in order to determine the "oil recovery" obtainable therefrom without any regard to scaling. Water floods have been discussed in general terms by Plummer and Woodward (1937), Plummer and Livingston (1940), Levine *et al.* (1941), Johnston and van Wingen (1945; also van Wingen and Johnston 1946), Yuster (1945, 1946; also Yuster and Calhoun 1945), Pfister and Breston (1947), Coomber and Tiratsoo (1950), Slobod and Caudle (1952), and Ryder (1949a, 1949b, 1949c). In particular, the influence of pressure (gradient) on water floods has been discussed by Ustinov (1938), Botset and Muskat (1939), Miller (1941), Calhoun *et al.* (1944), Morse and Yuster (1946, 1947), Ryder (1947), Breston and Hughes (1949), and Holmgren (1949). The influence of the rate of water injection on flooding experiments has been discussed by Earlougher (1943); the influence of the viscosity of the displacing fluid by Leverett (1939), Dykstra and Parsons (1948), and Everett *et al.* (1950). Similarly, the influence of "pore size" on water floods has been investigated by Uren and Fahmy (1927), Russell (1932), and Dale *et al.* (1950). The effect of the presence of a third phase, gas, in water flooding experiments of oil well cores has also been investigated, namely, by Breston and Hughes

(1948), and Schiffman and Breston (1950). Finally, Plummer *et al.* (1937), Torrey (1943), Dunning and Hsiao (1953), and Ojeda *et al.* (1953) discussed the effect of adding soap and other surface active agents to the flooding water, and Cloud (1941a, 1941b), Russell *et al.* (1947), and Sayre and Yuster (1949) investigated the influence of "connate water" already present in the oil cores.

Similar studies have been made by using air or gas "floods" (usually called "drives" in this case) in petroleum well cores originally saturated with liquid (oil or oil and water). Such studies have been reported by Ustinov (1938), Higgins (1940), McAllister (1941), Krutter (1941b), Hill and Guthrie (1943), Babson (1944), Day and Yuster (1944, 1945), Kaveler (1944), Arps (1945), Elkins (1946), Muskat and Taylor (1946b), Pirson (1946), Day (1947a, 1947b), Menzie *et al.* (1947; also Menzie and Nielsen, 1949), Nielsen and Menzie (1949), and by Patton (1947). One of the more general results of these experiments is (reported by Day and Yuster, 1945b; Muskat, 1946; and Nielsen and Menzie, 1948) that the oil saturation, starting with a completely saturated core, is a linear function of the logarithm of the volume of air injected to obtain that saturation. Stahl and Huntington (1943) investigated experimentally the influence of gravity.

Finally, experiments have also been reported where gas and water were used intermittently as displacing fluid (Nielsen and Yuster, 1945; Menzie *et al.*, 1946; Menzie, 1947, 1948; Pfister, 1947, Squires, 1947).

8.5. Physical Aspects of Relative Permeability

8.5.1. Visual Studies.
The extension of Darcy's law to multiple phase flow as performed in section 8.2.2. is, in fact, a heuristic procedure suggested by the analogy with single phase flow. It does not provide an understanding of the physics of multiple phase flow. However, as the law has been substantiated by experiments, it must be assumed that it is at least partly correct and a physical understanding of it has to be sought.

There are some direct investigations into the problem of double phase flow. Versluys (1917, 1931; see sec. 3) had already shown that, statically, two phases may be in three different states in a porous medium, which were termed as fully saturated, pendular, and funicular respectively. It must be assumed that the three states have different flow patterns.

If the porous medium is completely saturated with one phase, then a displacing phase must break a way into the displaced fluid. It is observed that this happens by establishing fingering into the porous medium. The fingers then branch out and eventually form a network.

This leads to a state where both the displacing and displaced fluid form a network of funicular channels where it is possible to get from any point in either fluid to any other point in the same fluid by not leaving that fluid. The displacing fluid will flow along the established network of its own (and so will the displaced fluid), gradually increasing its hold, until the channels of the displaced fluid break up and the latter then occurs in the pendular state. Eventually, depending on wettability characteristics, it may be possible that no more displaced fluid can be removed from the porous medium; the displacing fluid moves in its own network and the displaced phase may stay immobile as a "connate" phase.

The prevailing régime of double phase flow is that where both phases are flowing along their own funicular channels. It is easy to envisage that these channels may be much more tortuous than the pore channels for single phase flow. Thus, we arrive at a qualitative explanation for the low relative permeabilities characteristic of multiple phase flow systems.

A substantiation of these ideas has been achieved by visual (i.e., microscopic) flow studies. Such studies have been made by Mahoney (1947), Ryder (1948), Wilson and Calhoun (1952), Chatenever (1952; also Chatenever and Calhoun, 1952), and Muskat *et al.* (1953).

8.5.2. Theoretical Studies. Ultimately, it will be theoretical studies which will provide for the explanation of relative permeability. In qualitative terms, such studies were begun many years ago. Gardescu (1930) discussed the behaviour of gas bubbles in capillary spaces. Ceaglske and Kiesling (1940) studied the mechanics of capillary flow in solids, and Houpeurt (1949) investigated the effects of surface forces on the equilibrium and motion of oil deposits in soil.

A serious treatment of the theory of relative permeability will make use of the same concepts as those used in the treatment of single phase permeability. The first possibility, thus, is to establish heuristic correlations. This has been done by Atkinson (1948), Childs and Collis-George (1948), Harrington (1949), Whiting and Guerrero (1951), and Collis-George (1953).

The next possibility is to use capillaric models. Before doing so, however, one has to investigate the flow of several phases in a single capillary. This has been done by Bergelin (1949), Porkhaev (1949), and Templeton (1953, 1954). These investigations are still largely experimental, since techniques have been devised for the observation of gas-liquid or liquid-liquid displacements in uniform capillaries with diameters as small as four microns. Qualitative descriptions of the behaviour of moving interfaces were thus obtained. Quantitative observation of gas-liquid displacements with zero contact angle indicates the adequacy of a model based on Poiseuille's law, and the independence of capillary pressure from interfacial velocity.

The concepts of displacement in single capillaries have been applied to porous media, the latter represented as a bundle of such capillaries, by Gates and Lietz (1950), Fatt and Dykstra (1951), Burdine (1953), and Hassan and Nielsen (1953). A theory of relative permeability based upon such "capillaric models" has been given by Fatt and Dykstra (1951). It is outlined below.

As usual in capillaric model theories, it is assumed that the sample can be represented by a bundle of capillary tubes in which the fluid path length is

§8.5 PHYSICAL ASPECTS OF RELATIVE PERMEABILITY

not the same as the bulk length. In addition, the fluid path length varies with saturation. For a bundle of N capillary tubes the flow through dN tubes will be

$$dQ = q_{av}dN \tag{8.5.2.1}$$

where Q is the total flow rate through all the tubes and q_{av} is the average flow rate through the tubes in the interval dN. If the interval dN is made small, the average flow rate through a tube in the interval dN is given by Poiseuille's Law

$$q_{av} = \pi r^4 \Delta p / (8\mu l) \tag{8.5.2.2}$$

where Δp is the pressure drop across the tube of length l and radius r, and μ is the viscosity of the fluid. Substituting for q_{av} in equation (8.5.2.1), Fatt and Dykstra obtained

$$dQ = [\pi r^4 \Delta p / (8\mu l)]dN. \tag{8.5.2.3}$$

Darcy's law for linear flow of an incompressible fluid in a porous medium is

$$Q = kA\Delta p / (\mu L), \tag{8.5.2.4}$$

where k is the (total) permeability and Δp is the pressure drop across the bundle of tubes which has replaced the porous medium of length L and cross-sectional area A. Differentiating Darcy's law with A, Δp, μ and L constant, we obtain

$$dQ = (A\Delta p / (\mu L))dk. \tag{8.5.2.5}$$

Equating (8.5.2.3) and (8.5.2.5) we have

$$dk = \{\pi r^4 L / (8Al)\}dN. \tag{8.5.2.6}$$

If the pores are assumed to be cylinders,

$$dV = \pi r^2 l dN, \tag{8.5.2.7}$$

where V is the volume of flowing fluid in the pores. Substituting dN in equation (8.5.2.6) Fatt and Dykstra got

$$dk = \{r^2 L / (8Al^2)\}dV. \tag{8.5.2.8}$$

By definition, the saturation, s, of the porous sample is given by

$$s = V/V_p = V/(PAL), \tag{8.5.2.9}$$

where V_p is the total pore volume and P is the porosity. Differentiating equation (8.5.2.9) and substituting for dV in equation (8.5.2.8) we get

$$dk = ds r^2 P L^2 / (8l^2). \tag{8.5.2.10}$$

The ratio of the fluid path length, l, to the length L of the sample has earlier been termed "tortuosity," T:

$$T = l/L. \tag{8.5.2.11}$$

The tortuosity, T, has now to be related directly to measurable properties. Fatt and Dykstra did this by considering what happens to the wetting phase in the pore spaces as the sample is desaturated. During desaturation, the wetting phase retreats into the smaller pores and into the smallest crevices. Liquid in such crevices has a small radius of curvature and by the derivation given is considered to be in small pores. It can be reasoned then that liquid flowing in the crevices and small pores will travel a more tortuous path than liquid flowing through the large pores. As a first approximation, T can thus be assumed to vary inversely as r^b, (with $b > 0$)

$$T = a/r^b \qquad (8.5.2.12)$$

where a and b are constants. Substituting equations (8.5.2.11) and (8.5.2.12) in (8.5.2.10) gives

$$dk = \frac{Pr^{2(1+b)}}{8a^2} ds. \qquad (8.5.2.13)$$

The pressure, p_c, across a curved interface between two fluids is the capillary pressure given as follows:

$$p_c = 2\gamma \cos \theta / r, \qquad (8.5.2.14)$$

where r is the radius of curvature, γ the interfacial tension at the interface, and θ the liquid-solid contact angle. When equation (8.5.2.14) is applied to porous media, r is the "pore radius" at which a non-wetting phase just displaces a wetting phase out of the pore. Equation (8.5.2.14) thus gives r as a function of the capillary pressure. Substituting for r in equation (8.5.2.13) we obtain

$$dk = \frac{P(2\gamma \cos \theta)^{2(1+b)}}{8a^2 p_c^{2(1+b)}} ds. \qquad (8.5.2.15)$$

If the constants, a and b, were known, the effective permeability k_e, at any saturation, to the corresponding phase can be calculated by integrating equation (8.5.2.15) from zero saturation to the desired saturation s. Thus, Fatt and Dykstra obtained:

$$k_e = \frac{P(2\gamma \cos \theta)^{2(1+b)}}{8a^2} \int_0^s \frac{ds}{p_c^{2(1+b)}}. \qquad (8.5.2.16)$$

The relative permeability, k_r, however, can be calculated without knowing the constant a if the constant b is known or assumed, because by definition $k_r = k_e/k$, thus giving

$$k_r = \frac{\int_0^s ds/p_c^{2(1+b)}}{\int_0^1 ds/p_c^{2(1+b)}}. \qquad (8.5.2.17)$$

Gates and Lietz (1950) assumed $b = 0$, Fatt and Dykstra (1951), $b = \frac{1}{2}$. However, comparison of measured and calculated relative permeability data indicates that this constant may not be the same for all types of materials. In this instance, the constant must be held to be one of the usual undetermined factors so common in hydraulic radius theories.

The capillaric model theory has been improved by Purcell (1950) who attempted to calculate relative permeabilities by assuming pores which were doughnut-shaped instead of cylindrical.

A further refinement of the above theory is obtained if, instead of capillaric models, the Kozeny theory is extended to multiple phase flow. This has been done by Rose and co-workers (Rose and Bruce, 1949; Rose, 1949; Rose and Wyllie, 1949), Wyllie and co-workers (Wyllie, 1951; Wyllie and Spangler, 1952) and by Rapoport and Leas (1951).

Following Rose (1949), the Kozeny equation for single phase flow can be expressed as:
$$k = P/(S_p{}^2 T) \tag{8.5.2.17}$$

where P is the fractional porosity, S_p is the specific internal surface area of the pores per unit *pore* volume, and T is again a dimensionless textural constant related to the shape and orientation (tortuosity) of the pores. Carman (1941) later, in studying the properties of unconsolidated sands, claimed empirically that
$$S_p = p_D/\gamma \tag{8.5.2.18}$$

where p_D is the displacement pressure, and γ is the interfacial tension. Equations (8.5.2.17) and (8.5.2.18) can be combined to yield

$$k = \frac{P\gamma^2}{p_D{}^2 T}. \tag{8.5.2.19}$$

Now, considering the effective permeability to the wetting phase (k_{ew}) in a polyphase flow system, it can be argued by analogy with equation (8.5.2.19) that

$$k_{ew} = \frac{P_{ew}\gamma^2}{p_c{}^2 T_{ew}}, \tag{8.5.2.20}$$

where the effective porosity P_{ew} is equal to $s_w P$; and where p_c, the capillary pressure, is accepted as being the effective displacement pressure characterizing the partially saturated system. The effective textural constant, T_{ew}, is an undetermined factor and may therefore, for the purposes of this analysis, be assumed equal to T throughout the saturation range of interest. In any event, simply dividing equation (8.5.2.20) by equation (8.5.2.19) yields an expression for wetting phase relative permeability, k_{rw}, namely:

$$k_{rw} = \frac{k_{ew}}{k} = \frac{P_{ew}}{P} \frac{p_D{}^2}{p_c{}^2} \frac{T}{T_{ew}}, \tag{8.5.2.21}$$

which leads to the following equation, since $s_w = P_{ew}/P$:

$$k_{ew} = s_w \frac{p_D{}^2}{p_c{}^2}. \tag{8.5.2.22}$$

In order to test equation (8.5.2.22), it is necessary to have such pressure *versus* saturation data available as are obtained in capillary pressure (static) experiments or displacement (dynamic) experiments. Rose showed, however, that a more general relationship between the wetting phase relative permeability and saturation can be derived in which the relative permeability is entirely described in terms of explicitly known saturation parameters. This can be accomplished by treating the Kozeny equation (8.5.2.17) in a manner similar to the derivation of equation (8.5.2.22) from equation (8.5.2.19); that is, by assuming $T = T_{ew}$ through the saturation range of interest, and by recognizing that: $P_{ew}/P = s_{ew}$. Thus by analogy with equation (8.5.2.17), Rose set:

$$k_{ew} = \frac{P_{ew}}{S_{pew} T_{ew}} \tag{8.5.2.23}$$

or

$$k_{ew} = s_w \frac{S_p{}^2}{S_{pew}{}^2}. \tag{8.5.2.24}$$

This is the desired relationship. Here S_{pew} may be regarded as the effective surface separating the wetting phase from all other elements of the system. Comparison of equation (8.5.2.24) and equation (8.5.2.22) suggests that

$$S_{pew} = p_c/\gamma \tag{8.5.2.25}$$

is a reasonable expression for this effective surface area analogous to Carman's expression for specific surface area, equation (8.5.2.18).

An approach similar to that of Rose was chosen by Rapoport and Leas (1951) in order to give a theory of relative permeability. Following these authors, the total permeability is given by the Kozeny equation

$$k = \frac{cP^3}{S^2} \tag{8.5.2.26}$$

where S is now the specific surface area per unit *bulk* volume and c is the Kozeny constant, often set equal to $\frac{1}{5}$. In the case of double phase flow (phases "liquid" and "gas"), Rapoport and Leas represented the areas contacted by the two phases as follows:

$$\left. \begin{array}{l} S_L = R_L + I \quad \text{= specific surface area of the liquid system} \\ S_G = R_G + I \quad \text{= specific surface area of the gas system} \\ S = R_G + R_L \quad \text{= specific surface area of the solid,} \end{array} \right\} \tag{8.5.2.27}$$

where R_L represents the contact area between the solid and the liquid, R_G the contact area between the solid and the gas, and I the interfacial area between the two phases.

Referring to the conditions of applicability of Darcy's law, it is readily seen that the Kozeny equation can be applied to the liquid phase by (a) the replacement of the absolute porosity, P, by sP, the liquid filled pore volume per unit of bulk volume, i.e., by the "effective porosity" with respect to the liquid; (b) the substitution of the specific pore area, or total specific surface S by S_L, which represents the specific total boundary area of the liquid phase "flow matrix."

By substituting the above terms into equation (8.5.2.26), Rapoport and Leas obtained the effective permeability to the liquid phase as follows

$$k_L = \frac{cs^3 P^3}{S_L^2}, \qquad (8.5.2.28)$$

and by division of equation (8.5.2.28) by equation (8.5.2.27), they obtained the following expression for relative permeability to liquid:

$$k_L = s^3/(S_L/S)^2. \qquad (8.5.2.29)$$

In order to obtain a more usable expression for the relative permeability to liquid, the following considerations must be taken into account. It is generally recognized that towards the end of a capillary displacement process an irreducible minimum saturation s_0, is reached, and no more liquid can be forced out, except by diffusion. At this saturation the effective permeability to liquid becomes zero. Therefore, it seems logical to treat, in first approximation, the irreducible liquid as an essentially stationary element which reduces the porosity of the porous medium as well as the volume of flowing liquid. Once this liquid is considered as part of the solid framework, it is necessary to use an *effective* area, S_E, representing the surface that separates the flowing fluid from the immobile flow matrix composed of solid and stationary fluid, instead of the specific *solid* surface, S; furthermore, instead of using the total liquid saturation s, a *reduced saturation* s' has to be introduced:

$$s' = (s - s_0)/(1 - s_0). \qquad (8.5.2.30)$$

On substitution into equation (8.5.2.29), Rapoport and Leas obtained the following expression for the relative permeability to liquid:

$$k = \frac{s^3}{(S_L/S_E)^2} \left(\frac{s - s_0}{1 - s_0}\right)^3 \left(\frac{S_E}{S_L}\right)^2. \qquad (8.5.2.31)$$

It will be noted that the derivation of equation (8.5.2.31) implies the two following assumptions: (a) the molecular layers at the interfaces between solid liquid and gas liquid are stationary—if they were not, the basic Kozeny treatment should be slightly modified so as to take into account a certain "slippage" of the fluid on part of its boundaries; (b) the Kozeny constant of the wetting fluid "matrix" is the same as that of the total pore space—such an approximation seems to be justified since in general the possible variations of the Kozeny constant have been recognized to be comparatively small.

The calculation of k_L, the relative permeability to liquid, requires a knowledge of the surface areas, S_L and S_E. With the help of a thermodynamic approach the following expressions can be derived from a liquid gas capillary pressure curve:

$$S_G = \Sigma_s \equiv -(P/\gamma)\int_1^s p_c ds \tag{8.5.2.32}$$

$$S_E = \Sigma_{s_0} \equiv -(P/\gamma)\int_1^{s_0} p_c ds \tag{8.5.2.33}$$

$$S_L = \Sigma_w + 2I \equiv -(P/\gamma)\int_s^{s_0} p_c ds + 2I \tag{8.5.2.34}$$

in which the integrals indicate the areas under the p_c *versus* s curves measured between the indicated limits. It is seen that a system of three equations is obtained for the four unknowns S_G, S_E, S_L, and I. Consequently, while the values for S_G and S_E are directly indicated, separate solutions cannot be defined for S_L or I.

In order to evaluate S_L, Rapoport and Leas found it necessary to introduce an additional assumption concerning the distribution of the fluids. Such an assumption, expressed in terms of R_G, R_L, and I, will then represent one more independent equation, which together with the above relations will furnish a system from which all the unknowns can be deduced. It might be noted that on the basis of thermodynamic considerations only a more or less complex "trend" can be indicated for the liquid distribution, namely maximum contact area with the solid compatible with a minimum amount of interfacial area. In order to express such a trend in a more exact manner, it would be necessary to consider in detail the geometry of the porous medium which depends on many parameters such as grain size distribution, sphericity or shape factor of the grains, packing, consolidation, etc. Therefore, no exact additional relation can be established that would lead to unique solutions for S_L and I. The only logical approach that can be attempted consists of establishing the most general limiting conditions of fluid distribution that would be expected to apply to any, or at least to most, of the possible geometrical systems. According to such a principle, separate calculations for the minimum and for the maximum values of k_L can be made.

In accordance with the previously established relations, Rapoport and Leas considered the following system of equations:

$$\Sigma_w = S_L - 2I \tag{8.5.2.35}$$

$$\Sigma_s = R_G + I \tag{8.5.2.36}$$

where Σ_s and Σ_w have a well-defined value at any saturation and can, therefore, be considered as "constants" in regard to the unknowns S_L, I, and R_G. It is readily seen that, according to equation (8.5.2.35) for any set of Σ_s and Σ_w, i.e., for any saturation, a *maximum* value will be obtained for S_L if I is maximum, and that according to equation (8.5.2.16) the maximum value for I obtains if R_G is assumed to be zero. Consequently, Rapoport and Leas arrived at the following maximum value that S_L may take at any saturation:

$$S_L \text{ (Max.)} = \Sigma_w + 2\Sigma_s = \Sigma_{s_0} + \Sigma_s. \qquad (8.5.2.37)$$

According to equation (8.5.2.31) it is clear that at any saturation the smallest possible value will be obtained for the relative permeability to liquid k_L if a maximum area, S_L is considered, so that

$$k_L \text{ (Min.)} = s'^3 [S_E/S_L \text{ (Max.)}]^2 = s'^3 [\Sigma_{s_0}/(\Sigma_{s_0} + \Sigma_s)]^2. \qquad (8.5.2.38)$$

Physically the derivation of k_L (Min.) means that the distribution of the fluids is such that it does not permit the existence of any contact between the gas phase and the solid surface. The gas can then be visualized as flowing inside a network of channels completely surrounded by liquid and it might be noticed that such an assumption concerns only the disposition of the fluids, and does not correspond to any hypothesis about the geometry of the porous medium itself.

Referring to the system of equations (8.5.2.35) and (8.5.2.36), Rapoport and Leas showed that the theoretical absolute *minimum* value for S_L would obtain under the assumption $I = 0$, i.e., $S_L = \Sigma_w$. Such a situation would correspond to $R_G = S_G$, i.e., to the assumption of having the gas phase (and equally the liquid phase) bounded exclusively by solid surfaces. In that case, the porous medium should be visualized as formed by non-interconnected capillaries. It is clear that such an assumption is by far too limiting and leads actually to k_L values which are too high. Thus, instead of considering (even for the purpose of limiting conditions) a porous medium as equivalent to a bundle of capillary tubes, it is reasonable to represent it as an isotropic random packing of grains. Under this more general assumption, interfaces must exist between the gas and liquid, and none of the three surfaces areas, I, R_G, R_L is zero. The lowest possible value of the ratio I/R_G can then be evaluated. The use of this value in equation (8.5.2.36) leads to a minimum value of I, of S_L, and consequently, to maximum of k_L.

With the help of general capillary pressure relationships, it is possible to show that

$$I/R > (sP/(1-P))(p_m/p_c)^2 \qquad (8.5.2.39)$$

where p_c represents the capillary pressure at any saturation, s, and p_m the mean value of p_c corresponding to the saturation range s to s_0. The above inequality led Rapoport and Leas to postulate the following minimum value of S_L:

$$S \text{ (Min.)} = \Sigma_w + \frac{2\Sigma_{s_0}}{1 + \frac{1-P}{sP}\left(\frac{p_c}{p_m}\right)^2}, \qquad (8.5.2.40)$$

and the final expression for the limiting maximum relative permeability to liquid is then obtained as

$$k_L \text{ (Max.)} = s'^3 \frac{\Sigma_{s_0}^2}{\Sigma_w + \left\{\frac{2\Sigma_s}{1 + \frac{1-P}{sP}\left(\frac{p_c}{p_m}\right)^2}\right\}^2}. \qquad (8.5.2.41)$$

Equations (8.5.2.38) and (8.5.2.41) represent the minimum and maximum possible values of relative permeability, according to Rapoport and Leas.

The theories of relative permeability, outlined above, rely either on capillaric models or on the Kozeny theories. Therefore, they are subject to the usual limitations of the hydraulic radius theory. However, attempts to extend the drag theory or the statistical theory of hydrodynamics in porous media to double phase flow do not seem to have been attempted.

The papers on theory of relative permeability, mentioned above, also report experiments which were performed to substantiate their claims. However, as there are a variety of undetermined factors involved in those theories, an agreement of experiments and theory is not necessarily meaningful.

8.6. General Flow Equations for Immiscible Fluids

8.6.1. Limitations of Darcy's Law.
The limitations of Darcy's law in multiple phase flow of immiscible fluids are partly due to the same causes as in single phase flow. Thus, in the high velocity régime, one has to expect turbulence to play a role; and in the flow of gases, molecular effects and slippage phenomena might become important. Furthermore, there is also a series of "other" effects such as boundary phenomena, ionic phenomena, etc. In addition, there are effects due to hysteresis of contact angle as has been mentioned earlier.

The limitations of Darcy's relative permeability law have been discussed in general terms, for example, by van Wingen (1938), Muskat (1949), Ivakin (1951), and Richardson *et al.* (1952). In particular, boundary and related effects (such as macroscopic channelling) have been discussed by O'Connor (1946), Krynine (1950), Geffen *et al.* (1951), and Hill (1952).

Furthermore, there are some studies which claim that Darcy's law for multiple phase flow is not unconditionally valid, not even in low velocity régimes. Thus, Henderson and co-workers (Henderson and Yuster, 1948; Henderson and Meldrum, 1949) and Calhoun (1951b) found some dependence of relative permeability not only on saturation, but also on pressure. It is not clear, however, that this was not an experimental error. Furthermore, Yuster (1951) stated that relative permeabilities also depend on the absolute viscosities of the fluids involved. He tried to explain this by the remark that there is a shear transmitted at the two-phase interface within the porous medium which would actually entail such a phenomenon. A further discussion of this "Yuster effect" has also been given by Scott and Rose (1953). In addition, Sen-Gupta (1943) found abnormal changes in air permeabilities with water content in oil sands, and Geffen et al. (1951) observed hysteresis. These effects all signify limitations of Darcy's law.

8.6.2. Turbulence. It must be expected that turbulence plays a significant role in the high velocity régime of multiple phase flow in porous media. Unfortunately, analyses of turbulent effects are confined to the study of relatively coarse-grained porous media (pore diameters of the order of centimetres), such as are found in industrial towers. Turbulent flow of two immiscible phases in such towers has been investigated especially by chemical engineers. A great number of observations have been accumulated and various theories have been advanced to explain these observations.

The exposition of turbulent flow in coarse grained porous media, contained in this section (8.6.2), is essentially that given in a review article by Lerner and Grove (1951).*

Accordingly, most of the investigations on the behaviour of industrial towers are heuristic, corresponding to the heuristic investigations on single phase turbulent flow.

One of the earliest investigators of this problem, Peters (1922), observed that turbulent conditions of counter-current flow for a liquid-gas system occur rather abruptly. White (1934) extended this observation by noting that in a logarithmic plot of pressure drop *versus* gas velocity, two breaks ("critical" points) occur. With reference to the two breaks in the logarithmic plot, White defined the "loading point" of a column as "the gas velocity at which, for a given liquor rate, the logarithmic pressure drop-gas velocity curve first deviates from a slope of approximately two." The "flooding point" was defined as "that velocity at which the same curve turns abruptly almost vertically upward." This latter point was said to be accompanied by a marked spraying of the liquid. Further investigations along these lines were made by Simmons and Osborn (1934), Mayo et al. (1935), and Baker et al. (1935). The definition of loading point as given above was retained by Mach (1935) and again by White (1935).

Sherwood (1937) used the critical points of White's data interchangeably and changed the data of Baker et al. (1935) on loading velocities to flooding

---

* Permission to use the article by Lerner and Grove so extensively has kindly been granted by the American Chemical Society.

data. This may be readily seen by a comparison of Sherwood's work with the original data of Baker et al. A further anomaly arises from the fact that, although Baker et al. used the term "loading," they provided no definition of it.

Similar investigations were made by Uchida and Fujita (1936, 1937), Verschoor (1938), Vilbrandt et al. (1938), and Furnas and Bellinger (1938). Sherwood et al. (1938) revised their earlier work by using White's definitions for their own determinations of the loading and flooding points. Although they defined the flooding point as a graphical flood point, the flooding condition was actually taken "by visual observation of the liquid flowing over the packing and down the walls of the tower." Inasmuch as Sarchet (1942) has since reported an appreciable discrepancy between visual and graphical flood points, this would seem to render Sherwood's data inconsistent with his own definition, as the visual points were found by Sarchet to be from 15 to 20 per cent above or below the graphical flood points, the magnitude and direction of the deviation being a function of packing size. Sarchet further concluded that the graphical determination of the critical flow velocities was more dependable than visual observation. The basic correlation of flooding velocity used for design purposes in most instances is that of Sherwood, Shipley and Holloway (1938). While originally accurate only within 40 per cent, the correlation has been improved by further work on packing characteristics (see Lobo et al., 1945).

Considerably fewer data have been published on loading conditions than on flooding velocities. The data that have been published are somewhat contradictory. Elgin and Weiss (1939), measuring both *holdup* (i.e., *saturation*) and pressure drop, found no abrupt break point, but rather a gradual transition and suggested that the loading point be properly represented as a zone. On the other hand, Piret, Mann and Wall (1940) concluded from their data on a column 2.5 feet in diameter, packed with round gravel stones 1.75 inches in diameter, that a definite break in the holdup occurs at a point corresponding to the loading point. (See Lerner and Grove, 1951.)

Further experimental investigations with attempts at empirical correlations have been made by Cooper et al. (1941), Hendrix and Huntington (1941), Mertz and Huntington (1941), Jesser and Elgin (1943), Bain and Hougen (1944), Hassler et al. (1944), Martinelli et al. (1944, 1945), Brownell and Katz (1947), Zenz (1947), Ballard and Piret (1950), Breckenfeld and Wilke (1950), Hands et al. (1950), Kafarov and co-workers (Kafarov and Blyakhman, 1950; Kafarov and Planovskaya, 1951), Weisman and Bonilla (1950), Barth (1951), Crawford and Wilke (1951), Kocatas and Cornell (1954), and Sakiadis and Johnson (1954). Most of these investigations result in the plotting of some "correlation curves," but some of them end up in an analytical equation representing those curves. Particularly notable is the work of Brownell and Katz (1947) (see also Dombrowski and Brownell, 1954), who extended their correlation between friction factor and Reynolds number for single phase flow to double phase flow by including functions of saturation in the two correlated quantities.

It is difficult to judge how accurate the correlations advanced in the cited papers are if they are considered from a general standpoint. However, in view of the criticisms pointed out during the discussion of single phase flow correlations, it must be held that the present ones are subject to the same limitations. There seem to have been no fundamental improvements achieved over the basic methods used for single-phase flow correlations, which have been shown to be inadequate.

Of the theoretical attempts at correlations, Zenz (1947) has recently advanced a mechanism and a correlation for the limiting velocities of flow in packings based on an analogy with fluid flow through valves and orifices. Zenz smoothed the normal log-log plots into continuous curves, and by application of thermodynamic relationships defined the flood point as the gas velocity corresponding to a constant "critical pressure drop" above which the log-log pressure drop-gas velocity curves become vertical. The existence of a definite critical velocity for two-phase flow through orifices—i.e., a break point in the log-log plot—would obviate the basic premise of this latter theory (see Lerner and Grove, 1951).

Similar attempts at correlation based on the mechanisms of discontinuity were suggested by Sarchet (1942), Bertetti (1942), and Lerner and Grove (1951). The latter authors concluded that the variables under consideration, i.e., the superficial rates of flow based on the empty cross-sectional area of the bed, could not be directly related with the desired degree of accuracy. It was felt that, if the superficial rates could be converted to the actual flow rates through the interstices, then these latter variables would represent a better picture of the actual flow conditions, and would probably be more amenable to correlation.

In the case of countercurrent liquid-gas flow in vertical packed columns, qualitative analysis of the problem of converting superficial to actual flow rates leads directly to a general form of the quantitative relationships. If, as a first approximation, it is assumed that the liquid flows through the packing under a constant head, then any increase in the mass liquid throughput will result in an increase in the cross-sectional area through which the liquid flows, rather than an increase in the linear liquid velocity. It is implicitly assumed that the liquid-packing interaction forces remain constant, and that the velocity of the liquid is simply related to the free fall velocity under the given gravity head. Thus, if the increase in cross-sectional area through which the liquid flows is proportional to the increase in liquid rate, the mass velocity of the liquid phase remains substantially constant. On the other hand, the total flow area for both gas and liquid must remain constant so that the area pre-empted by the increase in liquid rate diminishes the area available for gas flow. Thus, the actual mass rate of flow of the gas increases with an increase in liquid rate, although the superficial gas flow rate is unchanged.

If now the limiting criterion for continuous gas flow be the velocity of the gas phase, it should be possible to achieve the loading and flooding conditions by an increase in liquid rate alone, at constant gas rate. Evidence that the actual gas flow rate is the limiting variable has been advanced by Furnas and Bellinger (1938), Piret *et al.* (1940), and Elgin and Weiss (1939), who found that the liquid saturation is independent of the gas rate up to the loading point, at which point it increases sharply. Furthermore, the fact that Elgin and Weiss were able to obtain flood points with zero gas flow is confirming evidence for the expectation of a point of discontinuity with increasing liquid rate.

The above reasoning may be placed on a quantitative basis by the use of the data of Jesser and Elgin (1943) on the relation between liquid rate and liquid saturation in packed columns. The objective of the analysis is to derive an expression relating what appears to be the critical variable, the actual gas flow rate, to the superficial liquid and gas rates.

Following the definition of Jesser and Elgin, the term "holdup," i.e., "liquid saturation," refers only to the dynamic holdup, which is that portion of the total saturation which varies with liquid rate. The static saturation, which is independent of liquid rate, is therefore included in the wet-drained fractional voids. Essentially this same procedure was utilized by Cooper *et al.* (1941) in their computation of the linear velocities of gas flow in packed columns.

Considering now a differential height in a vertical tower, we have from the definition of holdup (after Lerner and Grove, 1951):

$$A_L = a s_0 \tag{8.6.2.1}$$

where A_L is the cross-sectional area through which liquid is flowing, s_0 is the dynamic holdup (a fractional saturation), and a is a proportionality constant. The critical holdup (i.e., saturation) corresponding to the point of transition (flow discontinuity) at zero gas rate may be designated as $(s_0)_m$. Inasmuch as the critical flow point is marked by a break in the holdup, $(s_0)_m$ is not the holdup that completely occupies the free void space in the packing, but rather the holdup immediately preceding the attainment of the critical point at zero gas rate. In terms of relative velocity, zero gas throughput is obviously not equivalent to zero gas velocity relative to the falling liquid, but is nevertheless a reproducible reference flow for a given apparatus with a fixed free fall of liquid. Therefore, taking the zero gas throughput condition as the primary reference point, $(s_0)_m$ is a constant, and will result in liquid closure of the flow channel. Equation (8.6.2.1) then becomes:

$$A_T = a(s_0)_m \tag{8.6.2.2}$$

where A_T is the total area available for flow and may be defined as

$$A_T = A_g + A_L = P_{wd} A_0 \tag{8.6.2.3}$$

where A_g and A_L are the gas and liquid flow areas, respectively; A_0 is the cross-sectional area of the empty tower; and P_{wd} is the wetdrained fractional voids (a porosity value).

Equation (8.6.2.3) may be transposed to the form

$$A_g = A_T(1 - A_L/A_T). \tag{8.6.2.4}$$

Substituting equations (8.6.2.1) and (8.6.2.2) into (8.6.2.4), we obtain

$$A_g = A_T(1 - s_0/(s_0)_m) \tag{8.6.2.5}$$

or

$$A_g = P_{wd}A_0(1 - s_0/(s_0)_m). \tag{8.6.2.6}$$

By a simple material balance for the gas stream, the relation between the actual and superficial mass gas rates (denoted by G with appropriate subscripts) is given by

$$G_a = A_0 G_0/A_g \tag{8.6.2.7}$$

which is the Dupuit-Forchheimer assumption. Substituting equation (8.6.2.6) into equation (8.6.2.7), we obtain:

$$G_a = \frac{G_0}{P_{wd}(1 - s_0/(s_0)_m)}. \tag{8.6.2.8}$$

For the purpose of visualization it is advantageous to change from a mass flow rate to a pore velocity v_g:

$$v_g = \frac{G_0}{\rho_G P_{wd}(1 - s_0/(s_0)_m)}. \tag{8.6.2.9}$$

With water as the liquid, the data of Jesser and Elgin (1943) show

$$s_0 = bL_0^c \tag{8.6.2.10}$$

where b and c are constants for any given packing and L_0 is the superficial liquid mass rate. Substituting equation (8.6.2.10) in equation (8.6.2.9) and solving for G_0,

$$G_0 = v_g \rho_G P_{wd}(1 - bL_0^c/(s_0)_m). \tag{8.6.2.11}$$

Because $(s_0)_m$ is a constant, the superficial liquid and gas rates are now related through the actual pore velocity or, conversely, the pore velocity may be expressed as a function of L_0 and G_0.

Thus far, the actual mechanism of channel closure has not been considered. However, the literature contains considerable material on two-phase flow transition in pipes which is relevant to the problem of the mechanism of closure in packed column channels. Boelter and Kepner (1939), investigating two-phase flow in horizontal pipes, found that with increasing gas velocity, waves are generated on the surface of the liquid. As the gas velocity was further increased, the amplitude of the waves was observed to increase, until it became sufficiently large to cause the gas flow to become discontinuous. Although Boelter and Kepner employed parallel *horizontal* flow and an oil-air system, the same observations were made by O'Bannon (1924) and Houghten et al. (1924) for *vertical* and inclined countercurrent flow for air and water and steam and water. Recently, Martinelli and co-workers (1944, 1946) and Gazley (1949) have published extensive observations of the role of wave formation in channel closure.

The fact that a minimum gas velocity is necessary for inducing and sustaining waves on a liquid surface leads to the conclusion that the limiting velocity of two-phase flow is directly related to the gas velocity which is causing waveformation. For small channel diameters, of the order of magnitude of twice the gravitational wave amplitude, the limiting two-phase flow gas velocity and the critical velocity of wave formation may be assumed identical. Inasmuch as wave formation is primarily an interfacial phenomenon, the shape of the channel should have little effect on the inception velocity of the waves. For channel diameters larger than the critical minimum (twice the wave amplitude), the initiation of waves due to attainment of the critical gas velocity decreases the area available for gas flow and the velocity decreases so that an unstable situation results, terminating in channel closure. Therefore, the period between wave initiation and channel closure is necessarily short (Gazley, 1949). Within the limits of channel size ordinarily encountered in packed columns, the size and shape of the channel should have only a minor effect on the mechanism of closure, or the critical gas velocity of closure.

In view of the arguments cited, Lerner and Grove (1951) advanced the following postulates about flooding in packed columns in countercurrent two-phase flow:

(a) The mechanism of flooding in packed columns is wave formation on the liquid surface of sufficient amplitude to close the gas flow channel.

(b) Because the gas velocity appears to be the determining factor, there exists a definite velocity of gas flow at which waves will close the flow channel. This gas velocity is defined as the velocity of wave closure, $(v_g)_c$. One can also postulate that for any given liquid-gas system in countercurrent flow the velocity of wave closure is a constant, dependent primarily on the physical properties of the liquid and gas, and to a minor extent on the size and shape of the channel. On this basis, equation (8.6.2.11) becomes the correlating equation for limiting flow in packed columns.

Lerner and Grove (1951) substantiated their theory, outlined above, by a number of experiments. Although it is based upon semi-empirical arguments, and contains a number of undetermined factors, it seems to be at least qualitatively correct. It should not be forgotten, however, that it was essentially advanced in order to explain turbulent multiphase flow phenomena in rather *coarse* porous media such as are found in industrial towers. It should, therefore, not be expected to apply to substances with significantly finer pores.

As outlined earlier, systematic applications of either capillaric models or statistical mechanics to turbulent multiple phase flow phenomena, which would be applicable to media with fine pores, do not seem to have been undertaken.

8.6.3. Molecular Effects. As in single phase flow, the molecular effects can be split into two groups. First, if one of the phases is a gas, slippage will be expected; and second, if one of the phases is strongly adsorbed by the porous medium, adsorption and capillary condensation might play a role.

Gas slippage in double phase flow has been investigated by Rose (1948) and Fulton (1951). The essential results of these investigations are (i) that the value of relative gas permeability, extrapolated in the manner of Klinkenberg for infinite mean pressure, is the same as the non-wetting liquid relative permeability; and (ii) that the effect of gas slippage decreases as the wetting liquid saturation increases.

The multiple phase flow through adsorptive materials has been investigated essentially by soil-scientists, textile researchers, and building engineers, who are concerned with the motion of moisture through soil, textiles, or building materials, respectively; air being the second phase. It is generally assumed that the flow of moisture follows heuristically a diffusivity equation where the partial water vapour pressure is the driving agent. Experimental investigations of such flow phenomena have been made by Babbitt (1939, 1940), Christensen (1944), Peirce *et al.* (1945), Haugen and Marshall (1947), Hauser and McLaren (1948), Johansson *et al.* (1949), Wink and Dearth (1949), and Pfalzner (1950). Klute (1952) developed a numerical method for solving the flow equations. Kirkham and Feng (1949) formulated imbibition equations that contradict the usual diffusivity equation.

Understanding of diffusion of moisture through various materials is presumably to be sought in some application of the principles of adsorption. Babbitt (1948) observed that the permeability of hygroscopic materials decreases steadily at decreasing humidity, which he ascribes to the disappearance of sorbed layers. Accordingly, it seems that surface flow is the chief contributor to the total flow through the porous medium and thus that the phenomenon can be explained by an effect analogous to capillary condensation in the flow of condensable vapours.

8.7. Miscible Displacement

8.7.1. General Principles. Miscible displacement in porous media is a type of two-phase flow in which the two phases are completely soluble in each other. Therefore, capillary forces between the two fluids do not come into effect.

At first it might be thought that miscible displacement can be described in a very simple fashion. The mixture, under conditions of complete miscibility, could be thought to behave, locally at least, as a single phase fluid which would obey Darcy's law. The change of concentration, in turn, would be caused by diffusion along the flow channels and thus be governed by the bulk coefficient of diffusion of the one fluid in the other. In this fashion, one arrives at a heuristic description of miscible displacement which looks, at a first glance at least, very plausible.

In the case of incompressible fluids, the equations corresponding to the description given above, are as follows (Offeringa and Van der Poel, 1954):

(i) Darcy's law (cf. 4.3.1.2):

$$\boldsymbol{q} = -\frac{k}{\mu(C_1)}(\operatorname{grad} p - \rho(C_1)\boldsymbol{g}); \qquad (8.7.1.1)$$

(ii) Continuity-diffusivity condition:

$$-P\frac{\partial C_1}{\partial t} - \operatorname{div}(C_1\boldsymbol{q}) + PED_1 \operatorname{lap} C_1 = 0, \qquad (8.7.1.2)$$

where the indices 1, 2 refer to the two miscible phases, and C, the concentration, corresponds to the saturation s in immiscible displacement. The quantity D_1 is the usual diffusion coefficient of phase 1 in phase 2 in bulk masses of the liquids, and E is a dimensionless factor introduced to account for the fact that pore channels are tortuous and therefore longer than the corresponding macroscopic stream line intervals. Diffusion should take place along the tortuous channels and the factor E, therefore, should always be smaller than 1.

The flow equations (8.7.1.1/2) can be tested by experiment. If this is done, it soon becomes evident that they are inadequate to describe correctly the observed phenomena. This is seen, for instance, by devising scaling laws corresponding to the flow equations, and investigating whether systems which should be dynamically similar according to the theory, are actually dynamically similar in practice. Experiments to test this have been reported by Morse (1954), who has shown conclusively that scaling laws based on (8.7.1.1/2) do not obtain in practice.

Thus, according to experimental investigations, it appears that during miscible displacement, a fairly sharp concentration front becomes established, and that the front gets sharper the more slowly the displacement takes place. It also appears that below a finite displacement rate, a definite, special pattern of flow occurs. Thus it seems that a finite, characteristic time is necessary for the displacing fluid to diffuse sideways within the individual flow channels, thereby completely "mopping up" the original fluid. If the forward motion of the displacing fluid is slow, i.e., of the order of a pore diameter during the characteristic time, the displacement may be called *molecular*; if the displacement is fast (i.e., of the order of many pore diameters during the characteristic diffusion time), the displacement is *invasive*. In both cases, the degree of mixing seems governed by mechanical dispersion.

The phenomenological flow equations suggested by Darcy's law as a generalization of single phase flow to miscible multiple phase flow, do not allow for any characteristic times. A satisfactory heuristic theory of such phenomena does not yet exist. It is doubtful whether the heuristic approach will ever yield any suitable results in the present context, since it appears preferable to develop a theory of miscible displacement starting from microscopic and statistical considerations.

The above remarks pertain to what corresponds to the laminar flow régime in single phase flow. It must, of course, be expected that modifications have to be introduced in flow régimes where turbulence, molecular streaming, or other effects occur.

8.7.2. Specific Calculations. In practical applications of miscible displacement to engineering problems, the task is to calculate what happens during a miscible displacement process under specified external conditions.

As outlined above, there is really no satisfactory heuristic theory of miscible displacement in existence. It is therefore not indicated to try to solve equations (8.7.1.1/2) for practical cases owing to the latter's basic physical short-comings. Thus, efforts have been directed towards obtaining a simplified picture of the physical phenomena occurring during a displacement process and to apply the latter to practical cases.

From the general discussion in section 8.7.1, it becomes obvious that one can consider two extreme cases. If the displacement is very slow, each fluid behaves according to Darcy's law and there is very little mechanical mixing at the interface. If the displacement is fast, mechanical mixing plays a significant role.

A satisfactory approximation method for the calculation of practical cases exists only in the first of the above-mentioned extreme cases. In the case of such very slow displacement, mixing at the interface plays only a very minor role and it is possible to calculate what happens by using the same approximations as were used in the simplified theory of immiscible displacement (see sec. 8.3.2). Thus, each fluid may be assumed to obey Darcy's law (with its own viscosity) and the interface can be calculated by potential-theory methods. In fact, the use of such a simplified theory is probably much more justified in the present case of slow miscible displacement than in immiscible displacement for which it was originally invented, owing to the greater appropriateness of the approximations involved.

A beautiful confirmation of the validity of this type of approximation has been found in the behaviour of fresh water reservoirs beneath oceanic islands (cf. Wentworth, 1947, 1951). Thus it has been observed that rain water falling on the surface of say, a circular island seeps downward and accumulates at the surface of the salt water at sea level. The fresh water builds up to a height above sea level determined by the abundance of rain and the permeability of the island rock, thus forming a lens. At the lower boundary of the lens, the fresh water slowly displaces the sea water owing to the hydraulic head created by the build-up of water level at the top. Eventually a stationary state is reached wherein the lower boundary extends downward to a depth of about

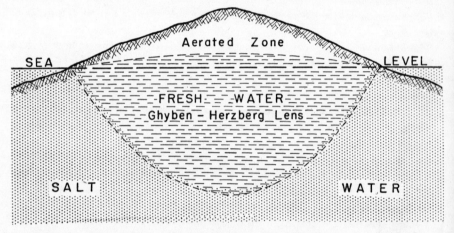

Figure 26. Drawing of a Ghyben-Herzberg lens beneath an oceanic island. (After Wentworth, 1947.)

forty times the height of fresh water above sea level, which corresponds to hydrostatic equilibrium of a body of fresh water floating upon sea water. The original displacement of sea water by fresh water must be assumed to have taken place over a very long period of time; the displacement was therefore almost without mechanical mixing. The eventual equilibrium, however, is not a hydrostatic one. If there were no fresh influx of rain water, the sea water would, eventually, diffuse into the fresh water. The appearance of a static equilibrium, therefore, is caused by a steady flow of fresh water into the sea, however, small this flux may be. The body of fresh water, therefore, has the shape of a doubly convex lens with the circular edge coinciding with the circular coast of the island. In groundwater hydrology, the principle of formation of a fresh water body beneath an island in the above manner is referred to as "Ghyben-Herzberg principle." A drawing of a Ghyben-Herzberg lens is shown in figure 26.

8.7.3. Microscopic Theories of Miscible Displacement. It has been stated in section 8.7.1 that the phenomenological theories of miscible displacement in porous media are inadequate. This fact is not really surprising in the light of earlier discussions of the physical aspects of permeability.

First, the factor E in (8.7.1.2) is tied up with the tortuosity of the pore channels, and it has been shown earlier that the concept of tortuosity is doubtful at best. Second, the diffusion of the one fluid into the other is certainly not taking place in the manner envisaged by the simple theory of section 8.7.1, but rather by a mechanism corresponding to Taylor's (sec. 2.5) theory of miscible displacement in tubes. Thus, the intrusion of the displacing fluid takes place along fingers down the centre of each flow channel, and diffusion, in turn, probably occurs sideways rather than forward. Third, it must be assumed that the intrusion of displacing fluid is greatly enhanced by the sideways dispersion of fluids caused by the random interconnections of the flow channels. This latter effect must be connected with the dispersivity of porous media introduced in the statistical theories of single phase flow.

To account for all of the above-mentioned phenomena during a miscible displacement process in porous media, is rather difficult, and only two attempts seem to have been made heretofore. The first is an extension of the idea of capillaric models, and the second is an extension of the statistical methods of section 6.5.3.

It is fairly straightforward to set up a capillaric model theory wherein Taylor's displacement process would obtain in each capillary. This has been done by Von Rosenberg (1956). Thus, if the porous medium be envisaged as a body consisting of straight parallel capillaries of uniform size and equal length, then exactly the same displacement law as was found by Taylor for each of the capillaries should hold. Thus, according to section 2.5, the "length" L of the saturation front in a linear displacement experiment should be directly proportional to the square root of the seepage velocity; and at a given velocity, the length L of the front should increase as the square root of the distance traversed. In mathematical symbols, one thus has for the length of the front:

$$L \sim \text{const.}\ t^{\frac{1}{2}} q \sim \text{const.}\ (qx)^{\frac{1}{2}} \qquad (8.7.3.1)$$

where $x \sim qt$; x is the mean linear position co-ordinate of the front, q the seepage velocity, and t is time.

The application of the statistical theory of single phase flow of section 6.5.3 can be extended to multiple phase flow if it is assumed that the displacing as well as the displaced fluid have similar physical characteristics. Then it is possible to identify the probability of a "particle" to be at a certain spot at a certain time (i.e., the quantity $v(x, t)$ in equation 6.5.3.26) with the "concentration" of the invading fluid within that originally contained in the porous medium. One can thus calculate what would happen in a linear displacement experiment where, at time "0," a displacing fluid is injected into the porous medium at

$x = 0$. Day (1956)* integrated equation (6.5.3.8) for the present case and found for the invading fluid concentration:

$$C = \int_z^{z=\infty} \frac{1}{\sqrt{2\pi}} e^{-\frac{1}{2}z^2} dz \qquad (8.7.3.2)$$

with

$$z = \frac{x - Vt}{2\sqrt{Dt}}. \qquad (8.7.3.3)$$

In interpreting this relation, it should be noted that D depends on q according to the assumption which is made regarding the microscopic flow law applicable to the individual capillaries of the porous medium, and according to the degree of molecular mixing within each flow channel. Assuming viscous flow, one has two limiting cases which yield (cf. sec. 6.5.3):

$$D = \text{const. } q^2, \qquad (8.7.3.4)$$

$$D = \text{const. } q. \qquad (8.7.3.5)$$

In the first case, the length of the saturation front is the same as that predicted by the application of Sir Geoffrey Taylor's theory (cf. equation 8.7.3.1), in the second case, it is independent of the velocity.†

It is thus seen that the statistical theory yields two limiting cases which could be thought of as related to the fact that there exists a characteristic time in miscible displacement. This, and the fact that experimentally the lengths of displacement fronts have been found to lie within the postulated limits, is a success of the statistical theory. However, it should not be forgotten that the present statistical theory is based upon an extremely simple model of flow through porous media. Discrepancies in further experimental tests with regard to the theory should therefore not be too unexpected.

* The writer is indebted to Dr. Day for an advance copy of his paper.

† The two limiting cases, thus, come out of the statistical theory described on pp. 113 ff. Originally, this theory was thought to yield only the limiting case represented by Eq. (8.7.3.4) and I am indebted to Dr. Todd of the University of California at Berkeley for having drawn my attention to the fact that there should be *two* limiting cases. In this instance, Dr. Todd reported investigations done by Mr. Rifai at Berkeley, based upon a different approach. However, it was the mention of these investigations of Rifai's which caused me to see that a similar conclusion can be drawn from the results of section 6.5.3.

LIST OF SYMBOLS*

a	radius of a tube; a constant; dispersivity
A	a constant; cross-sectional area
b	a constant
B	a constant
c	a constant; velocity of shock
C	concentration
C_p, C_v	specific heat at constant pressure or volume, respectively, of a gas
D	coefficient of diffusion (and related entities)
f	cumulative pore size distribution function; a force; a "function of"
F	Helmholtz free energy; area in the Kozeny theory; porosity factor
g	gravity acceleration
$\boldsymbol{g}$	vector of gravity acceleration (pointing downward)
G	a constant; mass rate of flow (for gas)
h	height of free surface above datum; height, width of a system
H	hydraulic head
$J(s)$	Leverett function of saturation
k	permeability
k_r	relative permeability
K	a constant
L	mass rate of flow for liquid; length of saturation front
lap	the Laplace operator
m	volume-to-surface ratio of capillary; a constant
$\boldsymbol{m}$	unit vector (components m_i)
M	molecular weight
n	denotes "a number"; ratio defined in a packing of spheres
$\boldsymbol{n}$	unit vector
N	counting number
p	fluid pressure
p_c	capillary pressure

* Note that vectors are printed in bold face type.

List of Symbols

P	porosity
$q, \mathbf{q}$	seepage velocity (vector)
Q	volume flow rate of a fluid
r	radial polar co-ordinate; fractional flow as a function of saturation
R	gas constant; radius
Re	Reynolds number
s	saturation; length of flow path
S	specific internal area (per unit bulk volume); total internal area
S_0	internal area per unit solid volume
S_p	internal area per unit pore volume
t	time
T	tortuosity
$\mathscr{T}, \mathscr{T}_{ik}$	overburden pressure in a porous medium
$\mathfrak{T}$	absolute temperature
v	local fluid velocity; probability function
V	volume; local fluid velocity
w	a probability function
W	work, energy per unit surface; equivalent pressure
x	a length
x, y, z	Cartesian co-ordinates (z vertical, upward)
y	adsorbate surface density
α	differential pore size distribution; a constant
β	relative pressure in adsorption experiments; a constant
β_f	compressibility of fluid
β_m	compressibility of porosity
γ	surface tension
δ	microscopic diameter such as pore diameter, etc.
Δ	symbol denoting finite difference
ε	proportionality factor
θ	contact angle
λ	mean free path length of molecules; friction factor
μ	viscosity
ρ	mass-density

List of Symbols

σ	a constant; standard deviation
Σ	interfacial area
φ	stream function; angular polar co-ordinate
ϕ	force-potential in fluid flow $= Hg$ if H is hydraulic head
Φ	various potential functions; "a function of"
χ	$\chi = \int_{p_0}^{p} \rho dp$
ψ	velocity potential in fluid flow
ω	number of adsorbed molecules per unit area
Ω	solid angle

BIBLIOGRAPHY*

INTRODUCTION

ENGELUND, F. (1953): On the laminar and turbulent flows of ground water through homogeneous sand. Trans. Danish Acad. Tech. Sci., no. 3, (105 pp.).
LEĬBENZON, L. S. (1947): Dvizhenie prirodnȳkh zhidkosteĭ i gazov v poristoĭ srede. Gosudarstv. izdat. tekh.-teoret. lit., Moscow and Leningrad (244 pp.).
MUSKAT, M. (1937): The flow of homogeneous fluids through porous media. 1st ed., 2nd prtg., publ. by Edwards, Ann Arbor, 1946 (763 pp.).
POLUBARINOVA-KOCHINA, P. YA. (1952): Teoriya dvizheniya gruntovȳkh vod. Gosudarstv. izdat. tekh.-teoret. lit., Moscow (676 pp.).
SCHEIDEGGER, A. E. (1956): Hydrodynamics in porous media. Encyclopedia of physics, vol. 9. Springer, Berlin.
ZIMENS, K. E. (1944): Poröse Stoffe, Kennzeichnung, Herstellung und Eigenschaften. Handbuch der Katalyse, vol. 4. Springer, Berlin (178 pp.).

PART I

ACKERMAN, A. S. E. (1945): Nature *155*, 82.
ARCHIE, G. E. (1942): Trans. AIME *146*, 54.
ATHY, L. F. (1930): Bull. Amer. Ass. Petrol. Geol. *14*, 1.
AVGUL, N. N. et al. (1951): Dokl. Akad. Nauk SSSR *76*, 855.
BABCOCK, A. B. (1945): Intern. Sugar J. *47*, 209.
BALLARD, J. H., PIRET, E. L. (1950): Ind. Eng. Chem. *42*, 1088.
BAL'SHIN, M. YU. (1949): Dokl. Akad. Nauk SSSR *67*, 831.
BEESON, C. M. (1950): Trans. AIME *189*, 313.
BELL, J. W. (1944): Trans. Can. Inst. Min. Met. *47*, 424.
BERNARD, R. A., WILHELM, R. H. (1950): Chem. Eng. Progr. *46*, 223.
BOND, R. L. et al. (1950): Fuel *29*, No. 4, 83.
BOTSET, H. G., REED, D. W. (1935): Bull. Amer. Ass. Petrol. Geol. *19*, 1053.
BRÖTZ, W., SPENGLER, H. (1950): Brennstoff-Chemie *31*, 97.
BRUSSET, H. (1948): C.R. Acad. Paris *227*, 843.
BUSBY, T. S. (1950): J. Soc. Glass Technol. *34*, 10.
CARMAN, P. C. (1938): J. Soc. Chem. Ind. *57*, 225.
CARPENTER, C. B., SPENCER, G. B. (1940): U.S. Bur. Mines R.I. 3540.
CHALKLEY, H. W. et al. (1949): Science *110*, 295.
CLOUD, W. F. (1941): Oil Weekly *103*, No. 8, 26.
CROSS, A. H. B., YOUNG, P. F. (1948): Trans. Brit. Ceram. Soc. *47*, 121.
DALLA VALLE, J. M. (1948): Micrometrics. Pitman, New York (555 pp.).
DALLMANN, H. (1941): Mühlenlab. *11*, 33.
DROTSCHMANN, C. (1943): Batterien *11*, 207.
DUMOND, J. W. M. (1947): Phys. Rev. *72*, 83.
EICHLER, B. (1950): J. Soc. Glass. Technol. *34*, 17.
FERRANDON, J. (1950): Ann. Inst. Tech. Bat. Trav. Publics No. *145*, 83.
FOORD, S. G. (1945): Nature *155*, 427.

* Slavonic names have been transliterated for this Bibliography according to the British (Cambridge) system; this is the system used, for instance, for *Physics Abstracts*.

FRANKLIN, A. D. et al. (1953): J. Appl. Phys. *24*, 1940.
FRICKE, H. (1931): Physics *1*, 106.
GASSMANN, F. (1951): Viertelj. Natf. Gesellsch. Zürich *96*, 1.
GILCHRIST, J. D., TAYLOR, J. (1951): J. Inst. Fuel *24*, 207.
GORODETSKIĬ, O. S. (1940): Ogneuporȳ *8*, 468.
GRACE, H. P. (1953): Chem. Eng. Progr. *49*, 303.
GRATON, L. C., FRASER, H. J. (1935): J. Geol. *43*, 785.
GRIFFITHS, J. C. (1952): Bull. Amer. Ass. Petrol. Geol. *36*, 205.
HALL, N. H. (1953): J. Petrol. Technol. *5*, sec. 1, 17.
HAWKSLEY, P. G. W. (1951): Bull. Brit. Coal Util. Res. Ass. *15*, 105.
HEYWOOD, H. (1938): Inst. Mech. Eng.; Nov. 1938; 52 pp.
———— (1947): Nature *159*, 717.
HOFSÄSS, M. (1948): Gas- und Wasserfach *89*, 139.
HRUBÍŠEK, J. (1941): Kolloid-Beihefte *53*, 385.
IMBT, W. C., ELLISON, S. P. (1946): A.P.I. Drill. Prod. Pract., 364.
KAYE, E., FREEMAN, M. F. (1949): World Oil *128*, No. 2, 24.
KIESSKALT, S., MATZ, G. (1951): Z. Ver. deut. Ing. *93*, No. 3, 58.
KING, J. G., WILKINS, E. T. (1944): Proc. Conf. Ultra-fine Struct. Coals and Cokes, Brit. Coal Util. Res. Ass., 46.
KISTLER, S. S. (1942): J. Phys. Chem. *46*, 19.
KNÖLL, H. (1940): Kolloid-Z. *90*, 189.
KRUMBEIN, W. C., PETTIJOHN, F. (1938): Manual of sedimentary petrography. Appleton-Century, London.
LAKIN, J. R. (1951): Trans. Brit. Ceram. Soc. *50*, 208.
LEĬBENZON, L. S. (1947): See Introduction.
LEPINGLE, M. (1945): Chaleur et ind. *26*, 101.
LOCKE, L. C., BLISS, J. E. (1950): World Oil *131*, No. 4, 206.
LOCKWOOD, W. N. (1950): Bull. Amer. Ass. Petrol. Geol. *34*, 2061.
LȲNOVSKIĬ, O. P., POSTNIKOVA, E. N. (1940): Voenno-Sanit., Nos. 8–9, 99.
MANEGOLD, E. (1937a): Kolloid-Z. *80*, 253.
———— (1937b): Kolloid-Z. *81*, 19.
———— (1941): Z. Ver. deut. Ing., Beih. Verfahrenstech., 44.
MANEGOLD, E., SOLF, K. (1939): Kolloid-Z. *89*, 36.
MCGEE, A. E. (1926): J. Amer. Ceram. Soc. *9*, 814.
MELCHER, A. F. (1921): Trans. AIME *65*, 469.
MILLS, G. L. (1948): Nature *161*, 313.
MILNER, H. B. (1940): Sedimentary petrography. T. Murby & Co., London.
MITTON, R. G. (1945): J. Int. Soc. Leather Trades' Chem. *29*, 255.
MÜLLER, H. (1947): Z. Naturf. *2a*, 473.
MUSKAT, M. (1937): See Introduction.
NISSAN, A. H. (1938): J. Inst. Petrol. Tech. *24*, 351.
NUSS, W. F., WHITING, R. L. (1947): Bull. Amer. Ass. Petrol. Geol. *31*, 2044.
OWEN, J. E. (1952): Trans. AIME *195*, 169.
PAGE, J. B. (1947): Proc. Soil Sci. Soc. Amer. *12*, 81.
PEERLKAMP, P. K. (1948): Landbouwkund. Tijdschr. *60*, 321.
POLLARD, T. A., REICHERTZ, P. P. (1952): Bull. Amer. Ass. Petrol. Geol. *36*, 230.
RALL, C. G., TALIAFERRO, D. B. (1949): Producers Monthly *13*, No. 11, 34.
RIM, M. (1952): Trans. Amer. Geophys, Un. *33*, 423.
RITTER, H. L., ERICH, L. C. (1948): Anal. Chem. *20*, 665.
ROSENFELD, M. A. (1949): Producers Monthly *13*, No. 7, 39.

RYDER, H. M. (1948): World Oil *127*, No. 13, 129.
SAUNDERS, H. L., TRESS, H. J. (1945): J. Iron Steel Inst., advance copy, July, 1945.
SCHOFIELD, R. K., TALIBUDDIN, O. (1948): Disc. Faraday Soc. *3*, 51.
SCHUMANN, H. (1944): Oel und Kohle *40*, 39.
SEIL, G. E. et al. (1940): J. Amer. Ceram. Soc. *23*, 330.
SHAPIRO, I., KOLTHOFF, M. (1948): J. Phys. Coll. Chem. *52*, 1020.
SHULL, C. G., et al. (1948): J. Amer. Chem. Soc. *70*, 1410.
SLICHTER, C. S. (1899): U.S. Geol. Surv. 19th Ann. Rept. Pt. 2, 295.
SMITH, W. O. et al. (1929): Phys. Rev. *34*, 1271.
STAMM, A. J. (1931): Physics *1*, 116.
STEINBERG, A. R. (1946): J. Soc. Chem. Ind. *65*, 314.
STEVENS, A. B. (1939): AIME TP. 1061.
STULL, R. T., JOHNSON, P. V. (1940): J. Res. Nat. Bur. Stand. *25*, 711.
TERZAGHI, K. (1951): Theoretical soil mechanics. Chapman and Hall Ltd., London.
TICKELL, F. G. et al. (1933): Trans. AIME *103*, 254.
TILLER, F. M. (1953); Chem. Eng. Progr, *49*, 467.
TOLLENAAR, D., BLOCKHUIS, G. (1950): Appl. Sci. Res. *A2*, 125.
UREN, L. C. (1943): Petrol. Eng. *14*, No. 9, 51.
VERBECK, G. J. (1947): J. Amer. Concrete Inst. *18*, 1025.
WALDO, A. W., YUSTER, S. T. (1937): Bull. Amer. Ass. Petrol. Geol. *21*, 259.
WASHBURN, W., BUNTING, E. N. (1922): J. Amer. Ceram. Soc. *5*, 112.
WESTMAN, A. E. R. (1926): J. Amer. Chem. Soc. *48*, 311.
WIGGINS, E. J. et al. (1939): Canad. J. Res. *B17*, 318.
WYLLIE, M. R. J., SPANGLER, M. B. (1952): Bull. Amer. Ass. Petrol. Geol. *36*, 359.
ZAVODOVSKAYA, E. K. (1937): Zavodsk. Lab. *6*, 1021.
ZIMENS, K. E. (1944): See Introduction.

PART II

ADAM, N. K. (1941): The physics and chemistry of surfaces. 3rd. ed., Oxford.
ADZUMI, H. (1937): Bull. Chem. Soc. Japan *12*, 292.
ARNELL, J. C. (1946): Canad. J. Res. *A24*, 103.
BROWN, R. C. (1947): Proc. Phys. Soc. London *59*, 429.
BRUNAUER, S. (1943): The adsorption of gases and vapors. Princeton Univ. Press, Princeton, N.J.
BURDEN, R. S. (1949): Surface tension and the spreading of liquids. Cambridge Univ. Press, London (92 pp.).
CASSIE, A. B. D. (1948): Disc. Faraday Soc. *3*, 11.
COMOLET, R. (1949): C.R. Acad. Paris *229*, 342.
DERYAGIN, B. V. et al. (1948): Dokl. Akad. Nauk SSSR *61*, 653.
ELTON, G. A. H. (1951): J. Chem. Phys. *19*, 1066.
EMMETT, P. H. (1948): Adv. in catalysis and rel, subj., *1*, 65.
GURNEY, C. (1949): Proc. Phys. Soc. London *A62*, 639.
HARTMAN, R. J. (1947): Colloid chemistry. 2nd ed., Riverside Press, Cambridge, Mass. (572 pp.).
JOOS, G. (1947): Theoretical physics. Blackie and Son, London (748 pp.).
KLOSE, W. (1931): Ann. Phys. *11*, 73.
KOENIG, F. (1950): J. Chem. Phys. *18*, 449.
KONCAR-DJURDEVIC, S. (1953): Nature *172*, 858.
KNUDSEN, M. (1909): Ann. Phys. (4) *28*, 75.
KUNDT, A., WARBURG, E. (1875); Pogg. Ann. Phys. *195*, 337 and 525,

LAMB, H. (1932): Hydrodynamics. 6th ed., Cambridge Univ. Press, London (738 pp.).
MACLELLAN, A. G. (1952): Proc. Roy. Soc. *A213*, 274.
MUSKAT, M. (1937): See Introduction.
PRIGOGINE, I. (1950): J. Chim. Phys. *47*, 33.
PRIGOGINE, I., MARÉCHAL, A. (1952): J. Colloid Sci. *7*, 122.
PRIGOGINE, I., SAROLÉA, L. (1952): J. Chim. Phys. *47*, 807.
SHUTTLEWORTH, R., BAILEY, G. L. (1948): Disc Faraday Soc. *3*, 16.
TAYLOR, (Sir) G. I. (1953): Proc. Roy. Soc. *A219*, 186.
TOPAKOGLU, C. (1951): Steady laminar flow of an incompressible fluid through a curved pipe of circular cross section (in Turkish). Pamphl. Istanbul Üniv. (30 pp.).
VON ROSENBERG, D. U. (1956): J. Amer. Inst. Chem. Eng. *2*, 55.

PART III

ADAM, N. K. (1948): Disc. Faraday Soc. *3*, 5.
ATTERBERG, A. (1918): Landw. Vers. Sta. *69*, 93.
BANGHAM, D. H. (1937): Trans. Faraday Soc. *33*, 805.
BARRETT, G. P., JOYNER, L. G. (1951): Anal. Chem. *23*, 791.
BARRETT, et al. (1951): J. Amer. Chem. Soc. *23*, 791.
BARTELL, F. E., OSTERHOF, H. J. (1927): Ind. Eng. Chem. *19*, 1277.
BOND, R. L., et al. (1948): Disc Faraday Soc. *3*, 29.
BOYER, R. L. et al. (1947): Trans. AIME *170*, 15.
BROAD, D. N., FOSTER, A. G. (1945): J. Chem. Soc. 366.
BROWN, H. W. (1951): Trans. AIME *192*, 67.
BRUCE, W. A., WELGE, H. J. (1947): Oil and Gas J. *46*, No. 12, 223.
BRUNAUER, S. (1943): See Part II.
BRUNAUER, S. et al. (1938): J. Amer. Chem. Soc. *60*, 309.
CALHOUN, J. C. et al. (1949) Trans. AIME *186*, 189.
CARMAN, P. C. (1941): Soil Sci. *52*, 1.
——— (1951): Proc. Roy. Soc. *A209*, 69.
CARMAN, P. C., RAAL, F. A. (1951): Proc. Roy. Soc. *A209*, 59.
CREMER, E. (1950): Z. Phys. Chem. *196*, 196.
DRAKE, L. C. (1949): Ind. Eng. Chem. *41*, 780.
DUNNING, H. N., et al. (1954): Petrol. Eng. *26*, No. 1, B82.
EMMETT, P. H. (1946): J. Amer. Chem. Soc. *68*, 1784.
——— (1948): See Part II.
EMMETT, P. H., BRUNAUER, S. (1937): Trans. Electrochem. Soc. *71*, 383.
EMMETT, P. H., DEWITT, T. W. (1943): J. Amer. Chem. Soc. *65*, 1253.
EVERETT, D. H. (1950): Trans. Faraday Soc. *46*, 453.
FOSTER, A. G. (1932): Trans. Faraday Soc. *18*, 645.
——— (1934): Proc. Roy. Soc. *A147*, 128.
——— (1945): J. Chem. Soc. *1945*, 769.
——— (1948): Disc. Faraday Soc. *3*, 41.
GLASSTONE, S. (1946): Textbook of physical chemistry, 2nd. ed., D. van Nostrand, New York (1320 pp.).
HACKETT, F. E., STRETTAN, J. S. (1928): J. Agr. Sci. *18*, 671.
HARKINS, W. D., JURA, G. (1943): J. Chem. Phys. *11*, 431.
——— (1944a): J. Amer. Chem. Soc. *66*, 919.
——— (1944b): J. Amer. Chem. Soc. *66*, 1362.
——— (1944c): J. Amer. Chem. Soc. *66*, 1366.
——— (1945): J. Chem. Phys. *13*, 449.

HARTMAN, R. J. (1947): Colloid chemistry. 2nd ed., Riverside Press, Cambridge, Mass. (572 pp.).
HASSLER, G. L., BRUNNER, E. (1945): AIME T.P. 1817 (10 pp.).
HENDERSON, L. M. et al. (1940): Refiner *19*, 185.
HIRST, W. (1947): Nature *159*, 267.
JOYNER, L. G. et al. (1945): J. Amer. Chem. Soc. *67*, 2182.
JOYNER, L. G. et al. (1951): J. Amer. Chem. Soc. *73*, 3155.
KATZ, S. M. (1949): J. Phys. Coll. Chem. *53*, 1166.
KEENAN, A. G. (1948): J. Chem. Educ. *25*, 666.
LAIRD, A. D. K., PUTNAM, J. A. (1951): Trans. AIME *192*, 275.
LANGMUIR, I. (1916): J. Amer. Chem. Soc. *38*, 2221.
LEVERETT, M. C. (1941): Trans. AIME *142*, 152.
LOISY, R. (1941): Bull. Soc. Chem. *8*, 589.
MCMILLAN, W. G., TELLER, E. (1951): J. Phys. Coll. Chem. *55*, 17.
MARTIN, M. et al. (1938): Geophysics *3*, 258.
MEYER, H. I. (1953): J. Appl. Phys. *24*, 510.
MORGAN, F. et al. (1950): Trans. AIME *189*, 183.
POLANYI, M. (1920): Z. Elektrochem. *26*, 370.
POLYAKOV, M. V. (1937): J. Phys. Chem. USSR *10*, 100.
POWERS, E. L., BOTSET, H. G. (1949): Producers Monthly *13*, No. 8, 15.
PURCELL, W. R. (1949): Trans. AIME *186*, 49.
RITTER, H. L., DRAKE, L. C. (1945): Ind. Eng. Chem., Anal. ed. *17*, 782.
ROSE, W. D., BRUCE, W. A. (1949): Trans. AIME *186*, 127.
SCHULTZE, K. (1925a): Kolloid-Z. *36*, 65.
——— (1925b): Kolloid-Z. *37*, 10.
SHULL, C. G. (1948): J. Amer. Chem. Soc. *70*, 1405.
SLOBOD, R. L. et al. (1951): Trans. AIME *192*, 127.
SMITH, W. O. et al. (1931): Physics *1*, 18.
STAHL, C. D., NIELSEN, R. F. (1950): Producers Monthly *14*, No. 3, 19.
THORNTON, O. F., MARSHALL, D. L. (1947): AIME T.P. 2126.
VERSLUYS, J. (1917): Int. Mitt. Bodenk. *7*, 117.
——— (1931): Bull. Amer. Ass. Petrol. Geol. *15*, 189.
WASHBURN, E. W. (1921): Proc. Nat. Acad. Sci. *7*, 115.
WENZEL, R. N. (1938): Ind. Eng. Chem. *28*, 988.
YUSTER, S. T., STAHL, C. D. (1948): Producers Monthly *13*, No. 2, 24.
ZSIGMONDY, R. (1911): Z. anorg. Chem. *71*, 356.

PART IV

AGADZHANOV, A. (1947): Hydrogeology and hydraulics of subterranean waters. Moscow.
ARAVIN, V. N. (1937): Trudȳ Leningrad industr. in-ta; Rasdel. gidrotekh. 1937, No. 19, 2.
ARONOVICI, V. S., DONNAN, W. W. (1946): Trans. Amer. Geophys. Un. *27*, 95.
BAVER, L. D. (1940): Soil physics. J. Wiley and Sons, New York.
BLANK, E. (1930): Handbuch der Bodenlehre. Springer, Berlin.
BROWN, R. L., BOLT, R. H. (1942): J. Acoust. Soc. Amer. *13*, 337.
CALHOUN, J. C. (1953): Fundamentals of reservoir engineering. Univ. Oklahoma Press, Norman (417 pp.).
CARMAN, P. C. (1938): See Part I.
DACHLER, R. (1936): Grundwasserströmung. Springer, Wien.

DALLA VALLE, J. M. (1943): See Part I.
DARCY, H. (1856): Les fontaines publiques de la ville de Dijon. Dalmont, Paris.
DICKEY, G. D., BRYDEN, C. L. (1946): Theory and practice of filtration. Reinhold, New York.
EASTMAN, J. E., CARLSON, A. J. (1940): AIME T.P. 1196.
FATT, I., DAVIS, D. H. (1952): Trans. AIME *195*, 329.
FERRANDON, J. (1948): Génie Civil *125*, 24.
FISHEL, V. C. (1935): Trans. Amer. Geophys. Un. *16*, 499.
FORCHHEIMER, P. (1930): Hydraulik. B. G. Teubner, Leipzig.
FROEHLICH, I. O. K. (1937): Mém. Ass. Int. Ponts Charpentes *5*, 133.
GARDNER, W. *et al.* (1934): Trans. Amer. Geophys. Un. *15*, 563.
GHIZETTI, A. (1949): Ann. Soc. Polon. Math. *22*, 195.
GRACE, H. P. (1953): Chem. Eng. Progr. *49*, 303.
GRIFFITHS, J. C. (1950): Producers Monthly *14*, No. 8, 26.
HANCOCK, R. T. (1942): Mining Mag. *67*, 179.
HUBBERT, M. K. (1940): J. Geol. *48*, 785.
IRMAY, S. (1951): Proc. Ass. Gén. Bruxelles; Ass. Int. Hydrol. (UGGI) *2*, 179.
IWANAMI, I. (1940): Trans. Soc. Mech. Eng. Japan *6*, No. 24, 4.
JACOB, C. E. (1940): Trans. Amer. Geophys. Un. *21*, 574.
JOHNSON, W. E., BRESTON, J. N. (1951): Producers Monthly, *15*, No. 4, 10.
JOHNSON, W. E., HUGHES, R. V. (1948): Producers Monthly *13*, No. 1, 17.
KAWAKAMI, M. (1933): J. Chem. Soc. Japan *54*, 133.
KEEN, B. A. (1931): Physical properties of the soil. Longmans Green, New York.
KOPPUIS, O. T., HOLTON, W. G. (1938): Phys. Rev. *51*, 684.
LEÏBENZON, L. S. (1946): See Introduction.
LEROSEN, A. L. (1942): J. Amer. Chem. Soc. *64*, 1905.
LEVA, M. *et al.* (1951): Flow through packed and fluidized systems. U.S. Bureau of Mines Bull. No. 504 (149 pp.).
LITWINISZYN, J. (1950): Ann. Soc. Polon. Math. *22*, 185.
MANEGOLD, E. (1938): Kolloid-Z. *82*, 269; *83*, 146; *83*, 299.
MEINZER, O. E. (1936): Bull. Amer. Ass. Petrol. Geol. *20*, 704.
MEINZER, O. E., FISHEL, V. C. (1934): Trans. Amer. Geophys. Un. *15*, 405.
MEINZER, O. E., WENZEL, L. K. (1940): Economic Geol. *35*, 915.
MITTON, R. G. (1945): See Part I.
MUSKAT, M. (1937): See Introduction.
—— (1949): Physical principles of oil production. McGraw-Hill, New York.
NEL'SON-SKORNYAKOV, F. B. (1949): Izd. Sovetsk. Nauka 1949 (58 pp.).
NEMENYI, P. (1933): Wasserbauliche Strömungslehre.
NUTTING, P. G. (1930): Bull. Amer. Ass. Petrol. Geol. *14*, 1337.
PALLMANN, H., DEUEL, H. (1945): Experientia *1*, 325.
PLUMMER, F. B., *et al.* (1934): AIME T.P. 578.
POLLARD, T. A., REICHERTZ, P.P. (1952): See Part I.
POLUBARINOVA-KOCHINA, P. YA. (1952): See Introduction.
PRESSLER, E. D. (1947): Bull. Amer. Ass. Petrol. Geol. *31*, 1851.
RIETEMA, K. (1951): Ingenieur *63*, ch. 1–ch. 7.
RUTH, B. F. (1946): Ind. Eng. Chem. *38*, 564.
RUTH, B. F. *et al.* (1933): Ind. Eng. Chem, *25*, 153.
SCHAFFERNAK, F. (1933): Wasserwirtsch. *1933*, 399.
SCHEIDEGGER, A. E. (1954): Geofis. Pura Appl. *28*, 75.
—— (1956): Geofis. Pura Appl. *33*, 111.

SCHWEIGL, K., FRITSCH, V. (1942): Gas- und Wasserfach *83*, 481 and 501.
SECCHI, I. M. (1936): Chimica e industr. *18*, 514 and 563.
SHCHELKACHEV, V. N. (1946a): C.R. Acad. Sci. URSS *52*, 103.
——— (1946b): C.R. Acad. Sci. URSS *52*, 203.
——— (1946c): C.R. Acad. Sci. URSS *52*, 392.
SILVERMAN, L. (1950): Quart. Ind. Hyg. Ass. *11*, 11.
SLATER, C. S. (1948): Agr. Eng. *29*, 119.
SOUERS, R. C., BINDER, R. C. (1952): Trans. Amer. Soc. Mech. Eng. *74*, 837.
STULL, R. T., JOHNSON, P. V. (1940): See Part I.
SULLIVAN, R. R. (1941): J. Appl. Phys. *12*, 503.
TERZAGHI, K. (1951): See Part I.
TILLER, F. M. (1953a): Pap. Sympos. Ind. Eng. Chem. Dec. 1953, Ann Arbor, Mich.
——— (1953b): Chem. Eng. Progr. *49*, No. 9, 467.
——— (1955): Chem. Eng. Progr. *50*, No. 6, 282.
TSUNKER, F. [Zunker] (1937): Vodnȳe svoĭstva pochv. Moscow.
VERSCHOOR, H. (1950a): Ingenieur Chem. Tech. *5* (June 2).
——— (1950b): Ingenieur *62*, 29.
VIBERT, A. (1939): Génie Civil *115*, 84.
VREEDENBURGH, C. G., STEVENS, O. (1936): Int. Conf. Soil Mech. Found. Eng. 1936.
WICKE, E., BRÖTZ, W. (1952): Chem. Ing.-Tech. *24*, 58.
WIGGINS, E. J., et al. (1939): See Part I.
WYCKOFF, R. D., et al. (1933): Rev. Sci. Instr. *4*, 394.

PART V

ARAVIN, V. I. (1936): Izv. NIIG *18*, 4.
——— (1937): Izv. NIIG *20*, 74.
——— (1940): Izv. NIIG *27*.
ARONOFSKY, J. S., FERRIS, O. D. (1954): J. Appl. Phys. *25*, 1289.
ARONOFSKY, J. S., JENKINS, R. (1952): Proc. 1st Nat. Congr. Appl. Mech., p. 763.
BARRON, R. A. (1948): Proc. 2nd Int. Conf. Soil Mech. Found. Eng. *3*, 209.
BAUMANN, P. (1951): Proc. Amer. Soc. Civ. Eng. *77*, separatum No. 86 (38 pp.).
BAZANOV, M. I. (1938): Prikl. Mat. Mekh, *2*, 223.
——— (1940): Izv. NIIG *28*, 116.
BIL'DYUG, E. I. (1940): Izv. NIIG *28*, 227.
BORELI, M. (1953): C.R. Acad. Sci. Paris *237*, 132.
BOULTON, N. S. (1942): Phil. Mag. *33*, 34.
——— (1951): J. Inst. Civ. Eng. *36*, No. 10, 534.
BOUSSINESQ, J. (1904): J. Math. *10*, 5 and 363.
BREITENÖDER, M. (1942): Ebene Grundwasserströmungen mit freier Oberfläche. Springer, Berlin.
BROWNSCOMBE, E. R., KERN, L. P. (1951): Petrol. Eng. *23*, No. 2, B7.
BRUCE, G. H., et al. (1953): Trans. AIME *198*, 79.
BRUCE, W. A. (1943): Trans. AIME *151*, 112.
CARSLAW, H. S., JAEGER, J. C. (1947): Conduction of heat in solids. Clarendon Press, Oxford.
CHARNIĬ, I. A. (1949): Izv. Akad. Nauk SSSR, Odt. tekh. nauk, 323.
——— (1950a): Izv. Akad. Nauk SSSR, Odt. tekh. nauk, 961.
——— (1950b): Izv. Akad. Nauk SSSR, Odt. tekh. nauk, 1326.

CHATAS, A. T. (1953a): Petrol. Eng. *25*, No. 5, B42.
——— (1953b): Petrol. Eng. *25*, No. 6, B38.
——— (1953c): Petrol. Eng. *25*, No. 9, B44.
CHILDS, E. C. (1947): Soil Sci. *63*, 361.
CHWALLA, K. (1948): Bautechnik *16*, 94 and 165.
CITRINI, D. (1951): Energia Elett. *28*, 1.
COOPER, H. H., JACOB, C. E. (1946): Trans. Amer. Geophys. Un. *27*, 526.
COURANT, R., HILBERT, D. (1943): Methoden der mathematischen Physik (2 vols.). Interscience Publ., New York.
DAVISON, B. B. (1936a): Phil. Mag. *21*, 881.
——— (1936b): Phil. Mag. *21*, 904.
DAVISON, B. B., ROSENHEAD, L. (1940): Proc. Roy. Soc. *A175*, 346.
DAY, P. R., LUTHIN, J. N. (1954): Proc. Soil Sci. Soc. Amer. *18*, 133.
DEVISON [DAVISON], B. B. (1937): Trudy gos. gidrol. in-ta, No. 5, 212.
DODSON, C. R., CARDWELL, W. T. (1945): Trans. AIME *160*, 56.
DOLCETTA, A. (1948): Energia Elett. *26*, 461.
DUPUIT, A. J. E. J. (1863): Etudes théoriques et pratiques sur le mouvement des eaux. Paris.
DYKSTRA, H., PARSONS, R. L. (1951): Trans. AIME *192*, 227.
EMERY, K. O., FOSTER, J. E. (1948): J. Marine Res. *7*, 644.
EVDOKIMOVA, V. A. (1950): Dokl. Akad. Nauk SSSR *74*, 669.
FANDEEV, V. V. (1947): Gidrotekh. Stroĭtel. 1947, No. 1.
FEYLESSOUFI, E. (1940): Bull. Tech. Suisse Rom. *66*, Nos. 19–20, 197.
FIL'CHAKOV, P. (1949): Dokl. Akad. Nauk SSSR *66*, 593.
——— (1951): Dokl. Akad. Nauk SSSR *78*, 41.
FORCHHEIMER, P. (1930): See Part IV.
GALIN, L. A. (1945): Dokl. Akad. Nauk SSSR *57*, 250.
——— (1951a): Prikl. Mat. Mekh. *15*, 111.
——— (1951b): Prikl. Mat. Mekh. *15*, 655.
GERSEVANOV, N. M. (1943): Izv. Akad. Nauk SSSR, Odt. tekh. nauk 73.
GIRINSKIĬ, N. K. (1937): Nauchnye zap. Mosk. in-ta inzh. vodnogo khoz, *1937* 4, 1.
——— (1938): Nauchnye zap. Mosk. gidro-melior. in-ta 1938, *5*, 47.
——— (1939): Nauchnye zap. Mosk. gidro-melior. in-ta 1939, *7*, No. 11, 3.
——— (1941): Stroĭizdat. *1941*.
——— (1946a): Dokl. Akad. Nauk. SSSR *54*, No. 3.
——— (1946b): Dokl. Akad. Nauk SSSR *51*, 337.
GOLUBEVA, O. V. (1950): Prikl. Mat. Mekh, *14*, 287.
GREEN, L., WILTS, C. H. (1952): Proc. 1st U.S. Nat. Congr. Appl. Mech., 777.
GUEVEL, P. (1953): C.R. Acad. Sci. Paris *237*, 597.
GÜNTHER, E. (1940): Forsch. Geb. Ingenieurw, *11*, 76.
HAMEL, G. (1934): Z. ang. Math. Mech. *14*, 129.
HANSEN, V. E. (1952a): Trans. Amer. Geophys. Un. *33*, 912.
——— (1952b): Proc. Amer. Soc. Civ. Eng. *78*, separatum No. 142 (18 pp.).
HETHERINGTON, C. R. *et al.* (1942): Trans. AIME *146*, 166.
HOPF, L., TREFFTZ, E. (1921): Z. ang. Math. Mech. *1*, 290.
HORNER, D. R. (1951): Proc. 3rd World Petrol. Congr. *2*, 503.
HUBBERT, M. K. (1940): J. Geol. *48*, 785.
HURST, W. (1934): Physics *5*, 20.
——— (1953): Petrol. Eng. *25*, No. 11, B6.

IRMAY, S. (1953): Proc. 3rd Int. Conf. Soil Mech. Found. Eng., 259.
IVAKIN, V. V. (1947): Izv. Akad. Nauk SSSR, Otd. tekh. nauk, 669.
JACOB, C. E. (1946): Trans. Amer. Geophys. Un. *27*, 198.
JACOB, C. E., LOHMANN, S. W. (1952): Trans. Amer. Geophys. Un. *33*, 559.
KALININ, N. K. (1943): Dokl. Akad. Nauk SSSR *30*, 393.
—— (1944): Dokl. Akad. Nauk SSSR *45*, 110.
—— (1948): Prikl. Mat. Mekh. *12*, 199.
—— (1950): Izv. Akad. Nauk SSSR, Odt. tekh. nauk, 1443.
—— (1952): Prikl. Mat. Mekh. *16*, 213.
KALININ, N. K., POLUBARINOVA-KOCHINA, P. YA. (1947): Prikl. Mat. Mekh. *11*, 231.
KAZARNOVASKAYA, B. É., POLUBARINOVA-KOCHINA, P. YA. (1943): Prikl. Mat. Mekh. *7*, 439.
KHOMOVSKAYA, E. D. (1937): Trudȳ gos. gidrolo. in-ta 1937, No. 5, 200.
KHRISTIANOVICH, S. A. (1941): Prikl. Mat. Mekh. *5*, 277.
KIRKHAM, D. (1939): Trans. Amer. Geophys. Un. *20*, 677.
—— (1940): Trans. Amer. Geophys. Un. *21*, 587.
—— (1945): Trans. Amer. Geophys. Un. *26*, 393.
—— (1949): Trans. Amer. Geophys. Un. *30*, 369.
—— (1950a): Trans. Amer. Geophys. Un. *31*, 425.
—— (1950b): J. Appl. Phys. *21*, 655.
KOCHINA, N. N. (1951): Prikl. Mat. Mekh. *15*, 679.
—— (1953): Prikl. Mat. Mekh. *17*, 377.
KOCHINA, N. N., POLUBARINOVA-KOCHINA, P. YA. (1952): Prikl. Mat. Mekh. *16*, 57.
KOZLOV, V. S. (1939): Izv. Akad. Nauk SSSR, Otd. tekh. nauk 89.
—— (1940a): Izv. Akad. Nauk SSSR, Otd. tekh. nauk, No. 3, 59.
—— (1940b): Izv. Akad. Nauk SSSR, Otd. tekh. nauk, No. 6, 58.
—— (1941): Dokl. Akad. Nauk SSSR *32*, 536.
KUFAREV, P. P. (1948): Dokl. Akad. Nauk SSSR *60*, 1333.
—— (1950a): Dokl. Akad. Nauk SSSR *75*, 353.
—— (1950b): Dokl. Akad. Nauk SSSR *75*, 507.
LAURENT, J. (1949): Rev. Gen. Hyd. *15*, No. 50, 79; No. 51, 143.
LEE, B. D. (1948): Trans. AIME *174*, 41.
LEFRANC, E. (1938): Génie Civil *113*, 162.
LEÏBENZON, L. S. (1945): Dokl. Akad. Nauk SSSR *47*, 15.
—— (1947): See Introduction.
LELLI, M. (1952): Energia Elett, *29*, 412.
LUKOMSKAYA, M. A. (1947): Prikl. Mat. Mekh. *11*, 621.
LUTHIN, J. N., GASKELL, R. E. (1950): Trans. Amer. Geophys. Un. *31*, 595.
LUTHIN, J. N., SCOTT, V. H. (1952): Agr. Eng. *33*, 279.
MACROBERTS, D. T. (1949): Trans. AIME *186*, 36.
MALCOR, R. (1948): Proc. 2nd Int. Conf. Soil Mech. Found. Eng. *7*, 169.
—— (1950): Travaux *34*, No. 183, 23.
MANDEL, J. (1939): Ann. Ponts Chaus. *109*, No. 7, 57.
—— (1951): Travaux *35*, 273.
MANSUR, C. I., PERRET, W. R. (1948): Proc. 2nd Int. Conf. Soil Mech. Found. Eng. *5*, 299.
MCNOWN, J. S., HSU, E. Y. (1950): Trans. Amer. Geophys. Un. *31*, 468.
MELESHCHENKO, N. T. (1936a): Izv. NIIG *18*, 44.
—— (1936b): Izv NIIG *19*, 25.
—— (1940): Izv. NIIG *28*, 234.

MERKEL, L. et al. (1949): Gas- und Wasserfach *90*, No. 9, 238.
MIKHAĬLOV, G. K. (1951): Dokl. Akad. Nauk SSSR *80*, 553.
────── (1953): Prikl. Mat. Mekh. *17*, 189.
MILLER, C. C. et al. (1950a): Trans. AIME *189*, 91.
MILLER, C. C. et al. (1950b): Petrol. Eng. *22*, No. 6, B7.
MORRIS, W. L. (1951): Ind. Eng. Chem. *43*, 2478.
MUSKAT, M. (1936): Trans. Amer. Geophys. Un. *17*, 391.
────── (1937): See Introduction.
────── (1943): Trans. AIME *151*, 175.
MUSKAT, M., BOTSET, H. G. (1931): Physics *1*, 27.
MYATIEV, A. N. (1946): Izv. Turkmen. Fil. Akad. Nauk SSSR, Nos. 3–4, 43.
NAHRGANG, G. (1954): Zur Theorie des vollkommenen und unvollkommenen Brunnens. Springer, Berlin.
NASBERG, V. M. (1951): Izv. Akad. Nauk SSSR, Otd. tekh. nauk, 1415.
NEDRIGA, V. P. (1947): Gidrotekh. Stroĭtel. *1947*, No. 5.
NEL'SON-SKORNYAKOV, F. B. (1937a): Nauchn. zap. Mosk. in-ta inzh. vodnogo khoz. 1937, No. 4, 77.
────── (1937b): Gidrotekh. Stroĭtel. No. 2, 32.
────── (1937c): Gidrotekh. Stroĭtel. Nos. 7–8, 25.
────── (1940a): Izv. Akad. Nauk SSSR, Otd. tekh. nauk, No. 6, 81.
────── (1940b): Izv. Akad. Nauk SSSR, Otd. tekh. nauk, No. 7, 53.
────── (1940c): Dokl. Akad. Nauk SSSR *28*, 483.
────── (1941a): Izv. Akad. Nauk SSSR, Otd. tekh. nauk, No. 1, 119.
────── (1941b): Izv. Akad. Nauk SSSR, Otd. tekh. nauk, No. 3, 39.
────── (1941c): Izv. Akad. Nauk SSSR, Otd. tekh. nauk, No. 1, 126.
────── (1947): Fil'tratsiya v odnorodnoĭ srede. Moscow.
NUMEROV, S. N. (1939): Izv. NIIG *25*.
────── (1940a): Izv. NIIG *27*.
────── (1940b): Izv. NIIG *28*, 64.
────── (1940c): Prikl. Math. Mekh. *4*, Nos. 5–6, 11.
────── (1940d): Izv. NIIG *28*, 91.
────── (1940e): Izv. NIIG *28*, 272.
────── (1942): Prikl. Mat. Mekh. *6*.
────── (1946): Izv. NIIG *31*, 273.
O'BRIEN, M. P., PUTNAM, J. A. (1941): AIME T.P. 1349.
PATTERSON, O. L. et al. (1951): A.P.I. Drill. Prod. Pract. 1951, 47.
PAVLOV, A. T. (1942): Prikl. Mat. Mekh. *6*, 281.
PAVLOVSKIĬ, N. N. (1936a): Izv. in-ta gidrotekh. *21*, 5.
────── (1936b): Izv. NIIG *19*, 49.
────── (1936c): Gidrotekh. Stroĭtel. Nos. 8–9.
PILATOVSKIĬ, V. P. (1953a): Prikl. Mat. Mekh. *17*, 179.
────── (1953b): Dokl. Akad. Nauk SSSR *89*, 635.
PISKUNOV, N. S. (1951): Dokl. Akad. Nauk SSSR *76*, 505.
PLUMMER, F. B., WOODWARD, J. S. (1936): Trans. AIME *123*, 120.
POLUBARINOVA-KOCHINA, P. YA. (1938): Izv. Akad. Nauk SSSR, Otd. mat. est. nauk; Ser. mat., No. 3, 371.
────── (1939a): Izv. Akad. Nauk SSSR, Ser. mat., No. 3, 329.
────── (1939b): Dokl. Akad. Nauk SSSR *24*, 1327.
────── (1939c): Izv. Akad. Nauk SSSR, Otd. tekh. nauk, No. 7, 45.
────── (1939d): Izv. Akad. Nauk SSSR, Ser. mat. Nos. 5–6, 579.

POLUBARINOVA-KOCHINA, P. YA. (1940a): Prikl. Mat. Mekh. *4*, 53.
―――― (1940b): Prikl. Mat. Mekh. *4*, 101.
―――― (1940c): Uchenye zap. Mosk. gos. in-ta 1940, Vȳp 39, mekh. 91.
―――― (1941): Prikl. Mat. Mekh. *5*, No. 2.
―――― (1942): Dokl. Akad. Nauk SSSR *34*, 46.
―――― (1945): Prikl. Mat. Mekh. *9*, 79.
―――― (1947): Prikl. Mat. Mekh. *11*, 357.
―――― (1948): Dokl. Akad. Nauk SSSR *63*, 623.
―――― (1950): Dokl. Akad. Nauk SSSR *75*, 511.
―――― (1951a): Prikl. Mat. Mekh. *15*, 511.
―――― (1951b): Prikl. Mat. Mekh. *15*, 649.
―――― (1952): See Introduction.
POLUBARINOVA-KOCHINA, P. YA., FAL'KOVICH, S. V. (1947): Prikl. Mat. Mekh. *11*, 629.
PȲKHACHEV, G. V. (1944): Trudȳ Grozn. neftyan. in-ta, Vȳp. 2.
RAM, G. *et al.* (1935): Proc. Ind. Acad. Sci. *2*, 22.
REINIUS, E. (1947): Tekn. Tidskr. *77*, 83.
RIZENKAMPF, B. K. (1938a): Uchenȳe zap. Saratovskogo gos. in-ta, Ser. fiz.-mat. *1* (14), Vȳp. 1, 89.
―――― (1938b): Uchenȳe zap. Saratovskogo gos. in-ta, Ser. mat.-mekh. *1* (14), Vȳp. 2, 181.
―――― (1940): Uchenȳe zap. Saratovskogo gos. in-ta, *14*, Vȳp. 5, 94.
ROBERTS, R. C. (1952): Proc. 1st U.S. Nat. Congr. Appl. Mech.
ROSSBACH, H. (1936–8): Ing. Arch. 1936, No. 1, 3, 5; 1938, No. 2.
ROSSBACH, H. (1941): Ing. Arch. *12*, 221.
SAUVAGE-DE-ST. MARC, F. (1947): Houille blanche *2*, 126.
SCHEIDEGGER, A. E. (1953): Petrol. Eng. *25*, No. 5, B121.
SEGAL, B. I. (1942): Dokl. Akad. Nauk SSSR *35*, 273.
―――― (1946): Izv. Akad. Nauk SSSR; Ser. mat., No. 10. 323.
SEGAL, B. I., GANTMAKHER, F. R. (1942): Dokl. Akad. Nauk SSSR *35*, 103.
SHAW, F. S., SOUTHWELL, R. V. (1941): Proc. Roy. Soc. *A178*, 1.
SHCHELKACHEV, V. N. (1946a): C.R. Acad. Sci. URSS *52*, 103.
―――― (1946b): C.R. Acad. Sci. URSS *52*, 203.
―――― (1946c): C.R. Acad. Sci. URSS *52*, 395.
―――― (1951): Dokl. Akad. Nauk SSSR *79*, 577.
SHCHELKACHEV, V. N., PȲKHACHEV, V. G. (1939): Interferentsiya skvazhin i teoriya plastovȳkh vodonapornȳkh sistem. Baku.
SHIMA, S. (1951): J. Soc. Civ. Eng. Japan *36*, No. 3, 3.
SOKOLOV, YU. D. (1951): Prikl. Mat. Mekh. *15*, 683.
―――― (1952): Ukrain. Mat. Zhur. *4*, 65.
TERRACINI, C. (1950): Atti Accad. Sci. Torino I. *84*, 139.
THEIS, C. V. (1934): Trans. Amer. Geophys. Un. *16*, 519.
TOPOLYANS'KIĬ, D. B. (1940): Dopovīdi Akad. Nauk *1940*, No. 7.
―――― (1947): Zbīr. prats' Inst. mat. Akad. Nauk URSR *1947*, No. 8.
UCHIDA, S. (1950): Rep. Inst. Sci. Technol. Tokyo *4*, Nos. 7–8, 200.
UGINCHUS, A. A. (1947): Gidrotekh. Stroĭtel., No. 5, 1.
VAN DEEMTER, J. J. (1949): Appl. Sci. Res. *A2*, No. 1, 33.
VAN EVERDINGEN, A. F., HURST, W. (1949): Trans. AIME *186*, 302.
VEDERNIKOV, V. V. (1939): Dokl. Akad. Nauk SSSR *23*, 335.
―――― (1940): Dokl. Akad. Nauk SSSR *28*, 408.

VEDERNIKOV, V. V. (1945): Dokl. Akad. Nauk SSSR *50*, 107.
────── (1947a): C.R. Acad. Sci. URSS *55*, 199.
────── (1947b): Gidrotekh. Stroĭtel., No. 1, 12.
────── (1948): Dokl. Akad. Nauk SSSR *59*, No. 6.
VERIGIN, N. N. (1940): Gidrotekh. Stroĭtel., No. 2, 30.
────── (1947): Gidrotekh. Stroĭtel., No. 5, 10.
────── (1949a): Dokl. Akad. Nauk SSSR *64*, 183.
────── (1949b): Izv. Akad. Nauk SSSR, Otd. tekh. nauk, 1723.
────── (1949c): Dokl. Akad. Nauk SSSR *66*, 589.
────── (1949d): Dokl. Akad. Nauk SSSR *66*, 1067.
────── (1950): Dokl. Akad. Nauk SSSR *70*, 777.
────── (1951): Dokl. Akad. Nauk SSSR *79*, 581.
────── (1953): Dokl. Akad. Nauk SSSR *89*, 627.
VIBERT, A. (1938a): Génie Civil *113*, 7.
────── (1938b): Génie Civil *113*, 406 and 427.
────── (1939): C.R. Acad. Paris *208*, 454.
VINOGRADOV, YU. P., KUFAREV, P. P. (1948): Prikl. Mat. Mekh. *12*, 181.
VLASOV, I. O. (1951): Prikl. Mat. Mekh. *15*, 117.
VOSHCHININ, A. P. (1939): Dokl. Akad. Nauk SSSR *25*, 133.
────── (1948): Prikl. Mat. Mekh. *12*, 761.
WERNER, P. W. (1946a): Trans. Amer. Geophys. Un. *27*, 687.
────── (1946b): J. Inst. Eng. Australia *18*, No. 6.
────── (1953): Geofis. Pura Appl. *25*, 37.
WERNER, P. W., NORÉN, D. (1951): Trans. Amer. Geophys. Un. *32*, 238.
WERNER, P. W., SUNDQUIST, K. J. (1951): Proc. Ass. Gén. Bruxelles, Ass. Int. Hydrol. (UGGI) *2*, 202.
WILSON, L. H., MILES, A. J. (1950): J. Appl. Phys. *21*, 532.
WYCKOFF, R. D., REED, D. W. (1935): Physics *6*, 395.
ZHUKOVSKIĬ, N. E. (1923): Poli. sobr. soch. ONTI (1937) *7*, 323.

PART VI

ADAMS, J. T. *et al.* (1949): Chem. Eng. Progr. *45*, 665.
ADAMSON, J. F. (1950): Nature *166*, 314.
ARAKAWA, M. *et al.* (1952): Bull. Inst. Chem. Res. Kyoto Univ. *29*, 78.
ARTHUR, J. R. *et al.* (1950): Trans. Faraday Soc. *46*, 270.
BACKER, S. (1951): Textile Res. J. *21*, 703.
BAKHMETEFF, B. A., FEODOROFF, N. V. (1937): J. Appl. Mech. *4A*, 97.
BARTELL, F. E., OSTERHOF, H. J. (1928): J. Phys. Chem. *32*, 1553.
BAVER, L. D. (1949): p. 364 in Meinzer's "Hydrology." Dover, New York.
BJERRUM, N., MANEGOLD, E. (1927): Kolloid-Z. *43*, 514.
BLAINE, R. L. (1941): Amer. Soc. Test. Mat. Bull. *108*, 17.
BLAINE, R. L., VALIS, H. J. (1949): J. Res. Nat. Bur. Stand. *42*, 257.
BREVDY, J. (1948): M. Sc. Thesis, Univ. of Minnesota, Minneapolis.
BRINKMAN, H. C. (1947): Appl. Sci. Res. *A1*, 27.
────── (1948): Appl. Sci. Res. *A1*, 81.
────── (1949): Research, Lond. *2*, 190.
BROOKS, C. S., PURCELL, W. R. (1952): Trans. AIME *195*, 289.
BROWN, J. C. (1950): TAPPI *33*, 130.
BUCHANAN, A. S., HEYMANN, E. (1948): Trans. Faraday Soc. *44*, 318.
BÜCHE, W. (1937): Z. Ver. deuts. Ing., Beih. Verfahrenstech, *1937*, 155.

BULNES, A. C., FITTING, R. V. (1945): Trans. AIME *160*, 179.
BURDINE, N. T. et al. (1950): Trans. AIME *189*, 195.
CALHOUN, J. C. (1953): See Part IV.
CARMAN, P. C. (1937): Trans. Inst. Chem. Eng. Lond. *15*, 150.
—— (1938a): Trans. Inst. Chem. Eng. Lond. *16*, 168.
—— (1938b): J. Soc. Chem. Ind. *57*, 225.
—— (1939a): J. Agr. Sci. *29*, 262.
—— (1939b): J. Soc. Chem. Ind. *58*, 1.
—— (1939c): J. Chem. Met. Mining Soc. S. Africa *39*, 266.
—— (1941): Soil Sci. *52*, 1.
—— (1948): Disc. Faraday Soc. 3, 72.
CARMAN, P. C., MALHERBE, P. LE R. (1950): J. Soc. Chem. Ind. 69, 134.
—— (1951): J. Appl. Chem. *1*, 105.
CHILDS, E. C., COLLIS-GEORGE, N. (1950): Proc. Roy. Soc. *A201*, 392.
CLOUD, W. F. (1941): Oil Weekly *103*, No. 8, 26.
COULSON, J. M. (1949): Trans. Inst. Chem. Eng. *27*, 237.
DALLA VALLE, T. M. (1938): Chem. and Met. Eng. *45*, 688.
DODD, C. G. et al. (1951): J. Phys. Coll. Chem. *55*, 684.
DONAT, J. (1929): Wasserkr. u. Wasserw. *24*, 225.
DUBININ, M. M. (1941): J. Appl. Chem. USSR *14*, 906.
ELTING, J. P., BARNES, J. C. (1948): Textile Res. J. *18*, 358.
EMERSLEBEN, O. (1924): Bautechnik 2, 73.
—— (1925): Physikal. Z. *26*, 601.
FAIR, G. M., HATCH, L. P. (1933): J. Amer. Water Wks. Ass. *25*, 1551.
FOWLER, J. L., HERTEL, K. L. (1940): J. Appl. Phys. *11*, 496.
FRANZINI, J. B. (1951): Trans. Amer. Geophys. Un. *32*, 443.
GOODEN, E. L., SMITH, C. M. (1940): Ind. Eng. Chem., An. Ed. *12*, 479.
GORING, D. A. I., MASON, S. G. (1950): Can. J. Res. *B28*, 307.
GRATON, L. C., FRASER, H. J. (1935): J. Geol. *43*, 785.
GRIFFITHS, J. C. (1950): Producers Monthly *14*, No. 8, 26.
—— (1952a): Bull. Amer. Ass. Petrol. Geol. *36*, 205.
—— (1952b): Penn. State Coll. M.I. Exp. Sta. Bull. *60*, 47.
GRIFFITHS, J. C., ROSENFELD, M. A. (1953): Amer. J. Sci. *251*, 192.
HEISS, J. F., COULL, J. (1952): Chem. Eng. Progr. *48*, 133.
HENDERSON, J. H. (1949): Producers Monthly *14*, No. 1, 32.
HEYWOOD, H. (1947): J. Imp. Coll. Eng. Chem. Soc. *3*, 7.
HOFFING, E. H., LOCKHART, F. J. (1951): Chem. Eng. Progr. *47*, No. 1, 3.
HUBBERT, M. K. (1940): J. Geol. *48*, 785.
HUDSON, H. E., ROBERTS, R. E. (1952): Proc. 2nd Midw. Conf. Fluid Mech. Eng. Series, *31*, No. 3, 105.
IBERALL, A. S. (1950): J. Res. Nat. Bur. Stand. *45*, 398.
JACOB, C. E. (1946a): Trans. Amer. Geophys. Un. *27*, 265.
—— (1946b): Trans. Amer. Geophys. Un. *27*, 245.
KAWAKAMI, M. (1932): J. Chem. Soc. Japan *53*, 1085.
KEYES, W. F. (1946): Ind. Eng. Chem., Anal. ed. *18*, 33.
KLYACHKO, V. A. (1948): Dokl. Akad. Nauk SSSR *60*, 1329.
KOLB, H. (1937): Zement *26*, 93.
KOZENY, J. (1927a): S.-Ber. Wiener Akad., Abt. IIa, *136*, 271.
—— (1927b): Wasserkr. u. Wasserw. *22*, 86.
—— (1932): Kulturtechniker *35*, 478.

Kraus, G., Ross, J. W. (1953): J. Phys. Chem. *57*, 334.
Kraus, G. *et al.* (1953): J. Phys. Chem. *57*, 330.
Krüger, E. (1918): Int. Mitt. Bodenk. *8*, 105.
Krumbein, W. C., Monk, G. D. (1942): Trans. AIME *151*, 153.
Kuhn, H. (1946): Experientia *2*, 64.
Kustov, B. I. (1949): Ogneupory̆ *14*, No. 6, 256.
Kwong, J. N. S. *et al.* (1949): Chem. Eng. Progr. *45*, 508.
Lamb, H. (1932): Hydrodynamics; 6th ed., Cambridge Univ. Press (738 pp.).
Lea, F. M., Nurse, R. W. (1939): J. Soc. Chem. Ind. Lond. *58*, T277.
Leïbenzon, L. S. (1947): See Introduction.
Loudon, A. G. (1952): Géotechnique Lond. *3*, 165.
Macey, H. H. (1940): Proc. Phys. Soc. *52*, 625.
Makhl, R. T. (1939): Keram. Sbornik No. 5, 45.
Manegold, E. (1937): Kolloid-Z. *81*, 164.
—— (1938): Kolloid-Z. *82*, 26.
Manegold, E., Solf, K. (1937): Kolloid-Z. *81*, 36.
—— (1938): Kolloid-Z. *82*, 135.
Martin, J. J. *et al.* (1951): Chem. Eng. Progr. *47*, No. 2, 91.
Mavis, F. T., Wilsey, E. F. (1937): Eng. News Rec. *118*, 299.
Missbach, A. (1938): Z. Zuckerind, Čekoslov. Rep. *62*.
Mitton, R. G. (1945): J. Int. Soc. Leather Trades' Chem. *29*, 255.
Mott, R. A. (1951): p. 242 in "Some aspects of fluid flow," Edward Arnold & Co. London.
Nelson, W. R., Baver, L. D. (1940): Proc. Soil Sci. Soc. Amer. *5*, 69.
Nemenyi, P. (1933): See Part IV.
Nutting, P. G. (1927): J. Franklin Inst., *203*, 313.
O'Neal, A. M. (1949): Soil Sci. *67*, 403.
Pecover, P. C. (1946): Commonwealth Eng. *33*, No. 12, 435.
Pfeiffenberger, G. W. (1946): Textile Res. J. *16*, 338.
Pillsbury, A. F. (1950): Soil Sci. *70*, 299.
Prockat, F. (1940): Chem. App. *27*, 129.
Purcell, W. R. (1949): Trans. AIME *186*, 39.
Rainard, L. W. (1947): Textile Res. J. *17*, 167.
Rigden, P. J. (1943): J. Soc. Chem. Ind. *62*, 1.
—— (1947): J. Soc. Chem. Ind. *66*, 130.
Robertson, A. A., Mason, S. G. (1949): Pulp & Paper Mag. Can. *50*, No. 13, 103.
Rose, H. E. (1945): Proc. Inst. Mech. Eng. Appl. Mech. *153*, No. 5, 141.
Rose, W. D., Bruce, W. A. (1949): Trans. AIME *186*, 127.
Scheidegger, A. E. (1953): Producers Monthly *17*, No. 10, 17.
—— (1954): J. Appl. Phys. *25*, 994.
—— (1956): Canad. J. Phys. *34*, 692.
Sen-Gupta, N. C., Nyun, M. G. T. (1943): Indian J. Phys. *17*, 39.
Shuster, W. W. (1952): Ph.D. Thesis, Rensselaer Polytech. Inst.
Slichter, C. S. (1899): See Part I.
Smith, W. O. (1932): Physics *3*, 139.
Sullivan, R. R. (1941): J. Appl. Phys. *12*, 503.
—— (1942): J. Appl. Phys. *13*, 725.
Sullivan, R. R., Hertel, K. L. (1940): J. Appl. Phys. *11*, 761.
—— (1942): in "Adv. in Coll. Sci." *1*, 37. Interscience, New York.
Svensson, J. (1949): Jernkontorets Annaler *133*, No. 2, 33.

TAUB, A. H. (1951): Proc. Midw. Conf. Fluid Dynamics (May 1950), 121.
TERZAGHI, C. (1925): Eng. News Rec. 95, 832.
THORNTON, O. F. (1949): Trans. AIME 186, 328.
TICKELL, F. G. (1935): Bull. Amer. Ass. Petrol. Geol. 19, 1233.
TICKELL, F. G., HIATT, W. N. (1938): Bull. Amer. Ass. Petrol. Geol. 22, 1272.
TICKELL, F. G. et al. (1933): Trans. AIME 103, 254.
TRAXLER, R. N., BAUM, L. A. H. (1936): Physics 7, 9.
WALAS, S. M. (1946): Trans. Amer. Inst. Chem. Eng. 42, 783.
WIGGIN, E. J. et al. (1939): Canad. J. Res. B17, 318.
WILSON, B. W. (1953): Austral. J. Appl. Sci. 4, 300.
WYLLIE, M. R. J. (1951): Trans. AIME 192, 1.
WYLLIE, M. R. J., GREGORY, A. R. (1955): Ind. Eng. Chem. 47, 1379.
WYLLIE, M. R. J., ROSE, W. D. (1950): Nature 165, 972.
WYLLIE, M. R. J., SPANGLER, M. B. (1952): Bull. Amer. Ass. Petrol. Geol. 36, 359.
YUHARA, K. (1954): Tikȳubuturi Geophys. Inst. Kȳoto Univ. 9, No. 2, 127.
ZHURAVLEVI, V. F., SȳCHEV, M. M. (1947): Zhur. Prikl. Khim. 20, No. 3, 171.
ZIMENS, K. E. (1944): See Introduction.
ZUNKER, F. (1932): Z. pfl. Ernähr. Düng. A25, 1.
——— (1933): Trans. 6th Comm. Int. Soil Sci. B18.

PART VII

ADZUMI, H. (1937a): Bull. Chem. Soc. Japan 12, 292.
——— (1937b): Bull. Chem. Soc. Japan 12, 304.
AEROV, M. E., UMNIK, N. N. (1950): Zhur. Prikl. Khim, 23, 1009.
ALLEN, H. V. (1944): Petrol. Refiner 23, No. 7, 247.
ARNELL, J. C. (1946): Canad. J. Res. A24, 103.
——— (1947): Canad. J. Res. A25, 191.
——— (1949): Canad. J. Res. A27, 207.
ARNELL, J. C., HENNEBERRY, G. (1948): Canad. J. Res. A26, 29.
ARONOFSKY, J. S. (1954): J. Appl. Phys. 25, 48.
ARTHUR, J. R. et al. (1950): Trans. Faraday Soc. 46, 270.
BAKHMETEFF, B. A., FEODOROFF, N. V. (1937): J. Appl. Mech. 4A, 97.
——— (1938): Proc. 5th Int. Congr. Appl. Mech., 555.
BARRER, R. M. (1939): Phil. Mag. 28, 148.
——— (1941): Diffusion in and through solids. Cambridge Univ. Press.
——— (1948): Disc. Faraday Soc. 3, 61.
——— (1949): Quart. Rev. Chem. Soc. London 3, 293.
BARRER, R. M., BARRIE, J. A. (1952): Proc. Roy. Soc. A213, 250.
BARRER, R. M., GROVE, D. M. (1951a): Trans. Faraday Soc. 47, 826.
——— (1951b): Trans. Faraday Soc. 47, 837.
BARTH, W., ESSER, W. (1933): Forschung 4, 82.
BERG, C. et al. (1949): Petrol. Refiner 28, No. 11, 113.
BIESEL, F. (1950): Houille blanche 5, 2, 157.
BLAKE, F. E. (1922): Trans. Amer. Inst. Chem. Eng. 14, 415.
BODMAN, G. B. (1937): Proc. Soil Sci. Soc. Amer. 2, 45.
BRESTON, J. N., JOHNSON, W. E. (1945): Producers Monthly 9, No. 12, 19.
BRIANT, R. C. (1945): Bull. Univ. Pittsburgh 41, No. 4, 38.
BRIDGWATER, A. B. (1950): Civ. Eng. Lond. 45, 234, 313, 385 451.
BRIEGHEL-MÜLLER, A. (1940): Kolloid-Z. 92, 285.
BRINKMAN, H. C. (1947): Appl. Sci. Res. A1, 27.

Brown, G. P. *et al.* (1946): J. Appl. Phys. *17*, 802.
Brown, R. L., Bolt, R. H. (1942): J. Acoust. Soc. Amer. *13*, 337.
Brownell, L. E., Katz, D. L. (1947): Chem. Eng. Progr. *43*, 537.
Brownell, L. E. *et al.* (1950): Chem. Eng. Progr. *46*, 415.
Bulkley, R. (1931): J. Res. Nat. Bur. Stand. *6*, 88.
Bull, H. B., Wronsky, J. P. (1937): J. Phys. Chem. *41*, 463.
Burke, S. P., Parry, V. F. (1935): AIME T.P. 607.
Burke, S. P., Plummer, W. B. (1928): Ind. Eng. Chem. *20*, 1196.
Calhoun, J. C. (1946): Ph.D. Thesis, Penn State Coll.
Calhoun, J. C., Yuster, S. T. (1946): A.P.I. Drill. Prod. Pract., 335.
Cambefort, H. (1951): Wwys. Exp. Sta. Transl. 51–3 (14 pp.).
Carman, P. C. (1937): Trans. Inst. Chem. Eng. *15*, 150.
—— (1947): Nature *160*, 301.
—— (1949): Nature *163*, 684.
—— (1950): Proc. Roy. Soc. *A203*, 55.
—— (1952): Proc. Roy. Soc. *A211*, 526.
Carman, P. C., Arnell, J. C. (1948): Canad. J. Res. *A26*, 128.
Carman, P. C., Malherbe, P. le R. (1950a): Proc. Roy. Soc. *A203*, 165.
—— (1950b): J. Soc. Chem. Ind. *69*, 134.
Carman, P. C., Raal, F. A. (1951): Proc. Roy. Soc. *A209*, 38.
Cassie, A. B. D. (1945): Trans. Faraday Soc. *41*, 458.
Chalmers, J. *et al.* (1932): Trans. AIME *98*, 375.
Chardabellas, P. E. (1940): Mitt. preuss. Vers. Anst. Wasser-, Erd- u. Schiffbau, H40.
Chilton, T. H. (1938): The science of petroleum. Oxford Univ. Press.
Chilton, T. H., Colburn, A. P. (1931a): Ind. Eng. Chem. *23*, 313.
—— (1931b): Ind. Eng. Chem. *23*, 913.
Collins, R. E. Crawford, P. B. (1953): J. Petrol. Techn. *5*, No. 12, Sec. 1, 19.
Cornell, D., Katz, D. L. (1951): Ind. Eng. Chem. *43*, 992.
—— (1953): Ind. Eng. Chem. *45*, 2145.
Danckwertz, P. V. (1953): Chem. Eng. Sci. *2*, 1.
Deryagin, B. V. (1946): Dokl. Akad. Nauk SSSR *53*, 627.
Deryagin, B. V., Krȳlov, N. A. (1944): Akad. Nauk SSSR, Otd. tekh. nauk, Inst. Mashinoved. Soveschanie po Vyazkosti Zhidkosteĭ i Kolloid Rastvorov *2*, 53.
Deryagin, B. V., *et al.* (1948): Dokl. Akad. Nauk SSSR *61*, 653.
Duriez, M. (1952): Ann. Trav. Pub. Belg. *104*, No. 2, 201.
Ehrenberger, R. (1928): Z. Oester. Ing. Arch. Ver., No. 9/10, 71.
Ekedahl, E., Sillén, L. G. (1947): Ark. Kemi. Min. Geol. *25A*, pap. 4 (26 pp.).
Elenbaas, J. R., Katz, D. L. (1948): Trans. AIME *174*, 25.
Engelund, F. (1953): Trans. Dan. Acad. Techn. Sci., No. 3, 105 pp.
Ergun, S. (1952a): Chem. Eng. Progr. *48*, 89.
—— (1952b): Analyt. Chem. *24*, 388.
—— (1953): Ind. Eng. Chem. *45*, 477.
Ergun, S., Orning, A. A. (1949): Ind. Eng. Chem. *41*, 1179.
Fair, G. M., Hatch, L. P. (1933): J. Amer. Water Wks. Ass. *25*, 1551.
Fancher, G. H., Lewis, J. A. (1933): Ind. Eng. Chem. *25*, 1139.
Fancher, G. H. *et al.* (1933): Penn. State M.I. Exp. Sta. Bull. *12*, 65.
Fancher, G. H. *et al.* (1934): Proc. World Petrol. Congr. *1*, 322.
Flood, E. A. *et al.* (1952a): Canad. J. Chem. *30*, 348.
—— (1952b): Canad. J. Chem. *30*, 372.

FLOOD, E. A. *et al.* (1952c): Canad. J. Chem. *30*, 380.
FORCHHEIMER, P. (1901): Z. Ver. deuts. Ing. *45*, 1782.
FOX, J. W. (1949): Proc. Phys. Soc. *B62*, 829.
FRIEDEL, F. A. (1949): Chem. Eng. Tech. *21*, 382.
FUJITA, S., UCHIDA, S. (1934): Soc. Chem. Ind. Japan *37*, 791B.
FURNAS, C. C. (1931): Ind. Eng. Chem. *23*, 1052.
GAMSON, B. W. *et al.* (1943): Trans. Amer. Inst. Chem. Eng. *39*, 1.
GIVAN, C. V. (1934): Trans. Amer. Geophys. Un. *15*, 572.
GLÜCKAUF, E. (1944): Nature *154*, 831.
GREEN, L., DUWEZ, P. (1951): J. Appl. Mech. *18*, 39.
GRIFFITHS, J. C. (1946): J. Inst. Petrol. *32*, 18.
GRIMLEY, S. S. (1945): Trans. Inst. Chem. Eng. *23*, 228.
GRISEL, F. (1936): C.R. Acad. Paris *203*, 1351.
GRUNBERG, L., NISSAN, A. H. (1943): J. Inst. Petrol. *29*, 193.
GUSTAFSON, Y. (1940): Lantbruks-Högskol. Ann. *8*, 425.
HANCOCK, R. T. (1942): Mining Mag. *67*, 179.
HAPPEL, J. (1949): Ind. Eng. Chem. *41*, 1161.
────── (1938): J. Appl. Mech. *5A*, 86.
────── (1940): J. Appl. Mech. *7*, 109.
────── (1943): Trans. Amer. Geophys. Un. *24*, 536.
HATFIELD, M. L. (1939): Ind. Eng. Chem. *31*, 1419.
HEID, J. G. *et al.* (1950): A.P.I. Drill. Prod. Pract., 230.
HICKOX, G. H. (1934): Trans. Amer. Geophys. Un. *15*, 567.
HILES, J., MOTT, R. A. (1945): Fuel *24*, 135.
HODGINS, J. W. *et al.* (1946): Canad. J. Res. *B24*, 167.
HOLLER, H. (1943): Deutsche Wasserw. *36*, 9.
HOLMES, W. R. (1946): Nature, *157*, 694.
HOOGSCHAGEN, J. (1953): J. Chem. Phys. *21*, 2097.
HUDSON, H. E., ROBERTS, R. E. (1952): See Part VI.
IBERALL, A. S. (1950): J. Res. Nat. Bur. Stand. *45*, 398.
ISHIKAWA, H. (1942): Waseda Appl. Chem. Soc. Bull. *19*, 51.
IWANAMI, S. (1940): Trans. Soc. Mech. Eng. Japan *6*, No. 24, 18.
JOHNSON, W., TALIAFERRO, D. B. (1938): U.S. Bur. Mines T.P. 592.
JONES, W. M. (1951): Trans. Faraday Soc. *47*, 381.
────── (1952): Trans. Faraday Soc. *48*, 562.
KEYES, W. F. (1946): Ind. Eng. Chem., An. Ed. *18*, 33.
KHANIN, A. A. (1948): Neftyanoe Khoz. *26*, No. 5, 32.
KLING, G. (1940): Z. Ver. deuts. Ing. *84*, 85.
KLINKENBERG, L. J. (1941): A.P.I. Drill. Prod. Pract., 200.
KRAUS, G., THIEM, J. R. (1950): J. Appl. Phys. *21*, 1065.
KRUTTER, H., DAY, R. J. (1941): Oil Weekly *104*, No. 4, 24.
KUZ'MINȲKH, I. N., APAKHOV, I. A. (1940): Org. Chem. Ind. USSR *7*, 257.
LAPUK, B. B., EVDOKIMOVA, V. A. (1951): Dokl. Akad. Nauk SSSR *76*, 509.
LEA, F. M., NURSE, R. W. (1947): Sympos. Particle Size Anal., Suppl. Trans. Inst. Chem. Eng. *25*, 54.
LEÏBENZON, L. S. (1945a): Izv. Akad. Nauk SSSR, Ser. Geog. Geofis. *9*, 3.
────── (1945b): Izv. Akad. Nauk SSSR, Ser. Geog. Geofiz. *9*, 7.
LEVA, M. (1947): Chem. Eng. Progr. *43*, 549.
────── (1949): Chem. Eng. *56*, No. 5, 115.
LEVA, M., GRUMMER, M. (1947): Chem. Eng. Progr. *43*, No. 11.

LEVA. M. *et al.* (1951): U.S. Bur. Mines Bull. No. 504 (149 pp.).
LINDQUIST, E. (1933): Proc. 1er Congr. Grands Barr. Stockholm, *5*, 81.
LINN, H. A. D. (1950): Water (Holland) *34*, 19.
LOCHMANN, G. (1940): Angew. Chemie *53*, 505.
MACH, E. (1935): Forschungsh. Ver. deuts. Ing. No. 375.
MACH, E. (1939): Dechema, Monographieen *6*, 38.
MANEGOLD, E. (1937): Kolloid-Z. *81*, 269.
MEINZER, O. E., WENZEL, L. K. (1940): Econom. Geol. *35*, No. 8, 915.
MEYER, G., WORK, L. T. (1937): Trans. Amer. Inst. Chem. Eng. *33*, 13.
MILLER, L. E. (1946): Producers Monthly *11*, No. 1, 35.
MILLER, K. T. *et al.* (1946): Producers Monthly *11*, No. 1, 31.
MISSBACH, A. (1937): Listy Cukrovar. *55*, 293.
MOODY, L. F. (1944): Trans. Amer. Soc. Mech. Eng. *66*, 671.
MORALES, M. *et al.* (1951): Ind. Eng. Chem. *43*, 225.
MORSE, R. D. (1949): Ind. Eng. Chem. *41*, 1117.
NAYAR, M. R., SHUKLA, K. P. (1943a): Current Sci. *12*, No. 5, 156.
―――― (1943b): Current Sci. *12*, No. 6, 183.
―――― (1943c): Current Sci. *12*, No. 7, 206.
―――― (1949): J. Sci. Ind. Res. *8B*, No. 8, 137.
NEMENYI, P. (1934): See Part IV.
NIELSEN, R. F. (1951): World Oil *132*, No. 6, 188.
NISSAN, A. H. (1942): J. Inst. Petroleum *28*, 257.
OMAN, A. O., WATSON, K. M. (1944): Natl. Petrol. News *36*, R795.
PENMAN, H. L. (1940a): J. Agr. Sci. *30*, 437.
―――― (1940b): J. Agr. Sci. *30*, 570.
PFALZNER, P. M. (1950): Canad. J. Res. *A28*, 389.
PLAIN, G. J., MORRISON, H. L. (1954): Amer. J. Phys. *22*, 143.
RADUSHKEVICH, L. V. (1941): J. Appl. Chem. USSR *14*, 900.
REYNOLDS, O. (1900): Papers on mechanical and physical subjects. Cambridge Univ. Press, London.
RIGDEN, P. J. (1946): Nature *157*, 268.
―――― (1947): J. Soc. Chem. Ind. Lond. *66*, T130.
ROMITA, P. L. (1951): Ricer. Scient. *21*, 1978.
ROSE, H. E. (1945a): Proc. Inst, Mech. Eng. *153*, 141.
―――― (1945b): Proc. Inst. Mech Eng. *153*, 148.
―――― (1945c): Proc. Inst. Mech. Eng. *153*, 154.
―――― (1949): Proc. Inst. Mech. Eng. *160*, 492.
―――― (1951): p. 136, in "Some aspects of fluid flow." Edward Arnold & Co., London.
ROSE, H. E., RIZK, A. M. A. (1949): Proc. Inst. Mech. Eng. *160*, 493.
ROSE, W. D. (1948): A.P.I. Drill. Prod. Pract., 209.
RUTH, B. F. (1946): Ind. Eng. Chem. *38*, 564.
SAMESHIMA, J. (1926): Bull. Chem. Soc. Japan *1*, 5.
ŠANDERA, K., MIRČEV, A. (1938): Listy Cukrovar *57*, 51.
SCHAFFERNAK, F., DACHLER, R. (1934): Wasserwirtsch. *1*, 145.
SCHEIDEGGER, A. E. (1953): Producers Monthly *17*, No. 10, 17.
―――― (1955): Geofis. Pura Appl. *30*, 17.
SCHOENBORN, E. M., DOUGHERTY, W. J. (1944): Trans. Amer. Inst. Chem. Eng. *40*, 51.
SCHWARTZ, C. E., SMITH, J. M. (1953): Ind. Eng. Chem. *45*, 1209.

SCHWERTZ, F. A. (1949): J. Appl. Phys. *20*, 1070.
SHOUMATOFF, N. (1952): Amer. Soc. Mech. Eng. Pap. No. 52, SA-42.
SHUKLA, K. P. (1944): Current Sci. *13*, 45.
SHUKLA, K. P., NAYAR, M. R. (1943): Current Sci. *12*, 155.
SILLÉN, L. G. (1946): Ark. Kemi. Min. Geol. *22A* (No. 5) Pap. 15 (22 pp.).
―― (1950a): Ark. Kemi *2*, Pap. 34, 477.
―― (1950b): Ark. Kemi *2*, Pap. 35, 499.
SILLÉN, L. G., EKEDAHL, E. (1946): Ark. Kemi Min. Geol. *22A* (No. 5) Pap. 16.
SOKOLOVSKIĬ, V. V. (1949a): Dokl. Akad. Nauk SSSR *65*, 617.
―― (1949b): Prikl. Mat. Mekh. *13*, 525.
SPAUGH, O. H. (1948): Food Technol. *2*, 33.
TAKAGI, S. (1947): J. Chem. Soc. Japan *68*, 5.
TAKAGI, M., ISHIKAWA, H. (1942): Waseda Appl. Chem. Soc. Bull. *19*, 59.
TAYLOR, G. I., DAVIES, R. M. (1947): Rep. Aero. Res. Counc. Lond. No. 2237, 14 pp.
TOMLINSON, R. H., FLOOD, E. A. (1948): Can. J. Res. *B26*, 38.
UCHIDA, S., FUJITA, S. J. (1934): Soc. Chem. Ind. Japan (Suppl. Binding) *37*, 724B.
UGUET, D. (1951): C.R. Acad. Paris *232*, 383.
URBAIN, P. (1941): C.R. Soc. Géol. France *1941*, 106.
VERONESE, A. (1941): L'ingenere *15*, 463.
VERSCHOOR, K. (1950a): Ingenieur (Utrecht) *62*, 29.
―― (1950b): Appl. Sci. Res. *A2*, 155.
VON ENGELHARDT, W., TUNN, W. L. M. (1954): Heidelberger Beiträge Min. Petrogr. *2*, 12.
WADELL, H. J. (1934): J. Franklin Inst. *217*, 459.
WARD, W. H. (1939): Engineering *148*, 435.
WEINTRAUB, M., LEVA, M. (1948): Chem. Eng. Progr. *44*, 801.
WENTWORTH, C. K. (1944): Amer. J. Sci. *242*, 478.
―― (1946): Trans. Amer. Geophys. Un. *27*, 540.
WHITE, A. M. (1935): Trans. Amer. Inst. Chem. Eng. *31*, 390.
WICKE, E. (1938): Z. Elektrochem. *44*, 587.
―― (1939a): Kolloid-Z. *86*, 167.
―― (1939b): Kolloid-Z. *86*, 295.
WICKE, E., KALLENBACH, R. (1941): Kolloid-Z. *97*, 135.
WICKE, E., TRAWINSKI, H. (1953): Chem. Ing. Tech. *25*, 114.
WICKE, E., VOIGT, U. (1947): Angew. Chem. *B19*, 94.
WILSON, L. H., et al. (1951): J. Appl. Phys. *22*, 1027.
WODNYANSZKY, G. (1938): Magyar Timár *2*, 33.
YUSTER, S. T. (1946): A.P.I. Drill. Prod. Pract., 356.
ZABEZHINSKIĬ, YA. L. (1939): J. Phys. Chem. USSR *13*, 1858.
ZEISBERG, F. C. (1919): Trans. Amer. Inst. Chem. Eng. *12*, 231.
ZHAVORONKOV, N. M. et al. (1949): Zhur. Fiz. Khim. *23*, 342.
ZUNKER, F. (1920): J. Gasbel. Wasservers.

PART VIII

ADAMSON, B., BROWN, B. D. (1953): J. Petrol. Technol. *5*, Sec. 1, 13.
ARPS, J. J. (1945): Trans. AIME *160*, 228.
ATKINSON, D. I. W. (1948): J. Imp. Coll. Chem. Eng. Soc. *4*, 78.
AVERYANOV, S. F. (1949): Dokl. Akad. Nauk SSSR *69*, 141.

BABBITT, J. D. (1939): Canad. J. Res. *A17*, 15.
—— (1940): Canad. J. Res. *A18*, 105.
—— (1948): Pulp Paper Mag. Canada *49*, 83.
BABSON, E. C. (1944): Trans. AIME *155*, 118.
BAIN, W. A., HOUGEN (1944): Trans. Amer. Inst. Chem. Eng. *40*, 29.
BAKER, T. et al. (1935): Trans. Amer. Inst. Chem. Eng. *31*, 296.
BALLARD, J. H., PIRET, E. L. (1950): Ind. Eng. Chem. *42*, 1088.
BARTELL, F. E., MILLER, F. L. (1932): Ind. Eng. Chem. *24*, 335.
BARTH, W. (1951): Chem. Ing. Tech. *23*, 289.
BERGELIN, O. P. (1949): Chem. Eng. *56*, No. 5. 104.
BERTETTI, J. W. (1942): Trans. Amer. Inst. Chem. Eng. *38*, 1023.
BOELTER, L. M. K., KEPNER, R. H. (1939): Ind. Eng. Chem. *31*, 426.
BOTSET, H. G. (1940): Trans. AIME *136*, 91.
BOTSET, H. G., MUSKAT, M. (1939): Trans. AIME *132*, 172.
BRAGINSKAYA, G. A. (1942): Prikl. Mat. Mekh. *6*.
BRANSON, U. S. (1951): World Oil *133*, No. 1, 184.
BRECKENFIELD, R. R., WILKE, C. R. (1950): Chem. Eng. Progr. *46*, No. 4, 187.
BRESTON, J. N., HUGHES, R. V. (1948): Producers Monthly *13*, No. 2, 14.
BRINKMAN, H. C. (1948): Appl. Sci. Res. *A1*, 333.
BROWNELL, L. E., KATZ, D. L. (1947): Chem. Eng. Progr. *43*, 601.
BROWNSCOMBE, E. R. et al. (1949): A.P.I. Drill. Prod. Pract., 302.
BROWNSCOMBE, E. R., DYES, A. B. (1952): A.P.I. Drill. Prod. Pract., 383.
BROWNSCOMBE, E. R. et al. (1950a): Oil and Gas J. *48*, No. 40. 68.
BROWNSCOMBE, E. R. et al. (1950b): Oil and Gas J. *48*, No. 41, 98.
BUCKLEY, S. E., LEVERETT, M. C. (1942): Trans. AIME *146*, 107.
BURDINE, N. T. (1953): Trans. AIME *198*, 71.
CALHOUN, J. C. (1951a): Oil and Gas J. *50*, No. 21, 117.
—— (1951b): Oil and Gas J. *50*, No. 20, 308.
—— (1953): See Part IV.
CALHOUN, J. C. et al. (1944): Producers Monthly *9*, No. 1, 12.
CARMAN, P. C. (1941): Soil Sci. *52*, 1.
CAUDLE, B. H. et al. (1951): Trans. AIME *192*, 145.
CEAGLSKE, N. H., KIESLING, F. C. (1940): Trans. AIME *36*, 211.
CHATENEVER, A. (1952): Oil and Gas J. *51*, No. 3, 174.
CHATENEVER, A., CALHOUN, J. C. (1952): Trans. AIME *195*, 149.
CHILDS, E. C. (1945): Soil Sci. *59*, 405.
CHILDS, E. C., COLLIS-GEORGE, N. (1948): Disc. Faraday Soc. *3*, 78.
CHRISTENSEN, H. R. (1944): Soil Sci. *57*, 381.
CHRISTIANSEN, J. E. (1944): Soil Sci. *58*, 355.
CLOUD, W. F. (1930): Trans. AIME *86*, 337.
—— (1941a): Oil Weekly *103*, No. 3, 33.
—— (1941b): Oil Weekly *103*, No. 6. 29.
COLLIS-GEORGE, N. (1953): Trans. Amer. Geophys. Un. *34*, 589.
COOMBER, S. E., TIRATSOO, E. N. (1950): J. Inst. Petrol. *36*, 543.
COOPER, C. M. et al. (1941): Trans. Amer. Inst. Chem. Eng. *37*, 979.
CRAWFORD, J. W., WILKE, C. R. (1951): Chem. Eng. Progr. *47*, 423.
DALE, C. B. et al. (1950): Producers Monthly. *14*, No. 7, 8.
DAY, P. R. (1956): Trans. Amer. Geophys. Un. *37* (in press).
DAY, R. J. (1947a): Producers Monthly *11*, No. 3, 16.
—— (1947b): World Oil *126*, No. 11, 36.

DAY, R. J., YUSTER, S. T. (1944): Producers Monthly *9*, No. 1.
—— (1945a): Oil Weekly *116*, No. 5, 24.
—— (1945b): Producers Monthly *10*, No. 1, 27.
DICKEY, P. A., ANDRESEN, K. H. (1945): A.P.I. Drill. Prod. Pract. 34.
DIETZ, D. N. (1953): Proc. k. Akad. Wet. *B56*, 83.
DOMBROWSKI, H. S., BROWNELL, L. E. (1954): Ind. Eng. Chem. *46*, 1207.
DUNLAP, E. N. (1938): Trans. AIME *127*, 215.
DUNNING, H. N., HSIAO, L. (1953): Bull. M.I. Exp. Sta. Penn, State Coll. *62*, 1.
DYKSTRA, H., PARSONS, R. L. (1948): A.P.I. Div. of Prod. Pap. No. 801–34K.
EARLOUGHER, R. C. (1943): Trans. AIME *151*, 125.
ELGIN, J. C., WEISS, F. B. (1939): Ind. Eng. Chem. *31*, 435.
ELKINS, L. E. (1946): A.P.I. Drill. Prod. Pract., 160.
ENRIGHT, R. J. (1954): Oil and Gas J. *53*, No. 2, 104.
EVERETT, J. P. et al. (1950): Trans. AIME *189*, 215.
EVINGER, H. H., MUSKAT, M. (1942): Trans. AIME *146*, 194.
FATT, I. (1953): J. Petrol. Technol. *5*, No. 10, 15.
FATT, I., DYKSTRA, H. (1951): Trans. AIME *192*, 249.
FLETCHER, J. E. (1949): Trans. Amer. Geophys. Un. *30*, 548.
FULTON, P. F. (1951): Producers Monthly *15*, No. 12, 14.
FURNAS, C. C., BELLINGER, F. (1938): Trans. Amer. Chem. Eng. *34*, 251.
GARDESCU, I. I. (1930): Trans. AIME *86*, 351.
GARRISON, A. D. (1934): A.P.I. Drill. Prod. Pract., 130.
GATES, J. I., LIETZ, W. T. (1950): A.P.I. Drill. Prod. Pract., 285.
GAZLEY, C. (1949): Ph.D. Thesis, Univ. of Delaware.
GEFFEN, T. M. et al. (1951): Trans. AIME *192*, 99.
HANDS, C. H. G. et al. (1950): J. Soc. Chem. Ind. *69*, 321.
HARRINGTON, J. D. (1949): M.Sc. Thesis, Univ. of Oklahoma.
HASSAN, M. E., NIELSEN, R. F. (1953): Petrol. Eng. *25*, No. 3, B61.
HASSLER, G. L. (1944): U.S. Patent 2,345,935.
HASSLER, G. L. et al. (1936): Trans. AIME *118*, 116.
HASSLER, G. L. et al. (1944): Trans. AIME *155*, 155.
HAUSER, P. M., MCLAREN, A. D. (1948): Ind. Eng. Chem. *40*, 112.
HENDERSON, J. H., MELDRUM, A. H. (1949): Producers Monthly *13*, No. 5, 12.
HENDERSON, J. H., YUSTER, S. T. (1948a): Producers Monthly *12*, No. 3, 13.
—— (1948b): World Oil *127*, No. 12, 139.
HENDERSON, J. H. et al. (1953): Trans. AIME *198*, 33.
HENDRIX, V. V., HUNTINGTON, R. L. (1941): Petrol. Eng. *13*, No. 1, 48.
HIGGINS, R. V. (1940): U.S. Bureau of Mines R.I. 3657.
HILL, H. B., GUTHRIE, R. K. (1943): U.S. Bureau of Mines R.I. 3715.
HILL, S. (1952): Chem. Eng. Sci. *1*, 247.
HOLMGREN, C. R. (1949): Trans. AIME *179*, 103.
HOLMGREN, C. R., MORSE, R. A. (1951): Trans. AIME *192*, 135.
HOUGEN, O. A., MARSHALL, W. R. (1947): Chem. Eng. Progr. *43*, No. 4.
HOUGHTON, F. C. et al. (1924): J. Amer. Soc. Heating Vent. Eng. *30*, 139.
HOUPEURT, A. (1949): Rév. Inst. franç. pétrole *4*, 107.
HUBBERT, M. K. (1940): See Part V.
—— (1950) J. Geology *58*, 655.
IVAKIN, V. V. (1951): Izv. Akad. Nauk SSSR. Otd. tekh. nauk *1951*, 1874.
JAMIN, J. C. (1860): C.R. Acad. Paris *50*, 172.
JESSER, B. W., ELGIN, J. C. (1943): Trans. Amer. Inst. Chem. Eng. *39*, 277.

JOHANSSON, C. H. et al. (1949): Acta Polytech. No. 29 (Chem. Met. Ser. No. 7), 5.
JOHNSTON, N., VAN WINGEN, N. (1945): A.P.I. Drill, Prod. Pract., 201.
JONES, P. J. (1946): Petroleum production. Reinhold, New York.
—— (1949): World Oil *129*, No. 2, 170.
KAFAROV, V. V., BLYAKHMAN, L. I. (1950): Zhur. Prikl. Khim. *23*, 244.
—— (1951): Zhur. Prikl. Khim. *24*, 624.
KAVELER, H. H. (1944): Trans. AIME *155*, 56.
KAZARNOVSKAYA, B. É. (1947): Dokl. Akad. Nauk SSSR *55*, 603.
KERN, L. R. (1952): Trans. AIME *195*, 39.
KINNEY, P. T., NIELSEN, R. F. (1949): Bull. M.I. Exp. Sta. Penn, State Coll. No. 54.
—— (1950): Producers Monthly, *14*, No. 3, 29.
KIRKHAM, D., FENG, C. L. (1949): Soil Sci. *67*. 29.
KLUTE, A. (1952): Soil Sci. *72*, 105
KOCATAS, B. M., CORNELL, D. (1954): Ind. Eng. Chem. *46*, 1219.
KRUTTER, H. (1941a): Producers Monthly *6*, No. 8, 25.
—— (1941b): Oil Weekly *102*, No. 1, 21.
KRUTTER, H., DAY, R. J. (1943): Pet. Chem. Techn. *6*, 1.
KRYNINE, D. P. (1950): Highway Res. Board, Proc. 29th Ann. Meet., 520.
LEAS, W. J. et al. (1950): Trans. AIME *189*, 65.
LERNER, B. J., GROVE, C. S. (1951): Ind. Eng. Chem. *43*, 216.
LEVERETT, M. C. (1939): Trans. AIME *142*, 152.
LEVERETT, M. C., LEWIS, W. B. (1941): Trans. AIME *142*, 107.
LEVINE, J. S. (1941): Oil Weekly *103*, No. 4, 34.
LEWIS, J. O. (1944): Trans. AIME *155*, 131.
LOBO, W. E. et al. (1949): Trans. Amer. Inst. Chem. Eng. *1949*, 693.
MACH, E. (1935): See Part VII.
MAHONEY, C. F. (1947): Thesis. Univ. of Oklahoma.
MARTINELLI, R. C. et al. (1944): Trans. Amer. Inst. Mech. Eng. *66*, 139.
MARTINELLI, R. C. et al. (1945): Trans. Amer. Inst. Chem. Eng. *41*.
MARTINELLI, R. C. et al. (1946): Trans. Amer. Inst. Chem. Eng. *42*, 681.
MAYO, F. et al. (1935): J. Soc. Chem. Ind. *54*, 373T.
MCALLISTER, E. W. (1941): Trans. AIME *142*, 39.
MENZIE, D. E. (1947): Producers Monthly *12*, No. 1, 28.
—— (1948): Oil and Gas J. *46*, No. 38, 78.
MENZIE, D. E., NIELSEN, R. F. (1949): Producers Monthly *13*, No. 5, 18.
MENZIE, D. E. et al. (1946): Producers Monthly *11*, No. 2, 14.
MENZIE, D. E. et al. (1947): Oil Weekly *125*, No. 9. 34.
MERTZ, R. V., HUNTINGTON, R. L. (1941): Oil Weekly *103*, No. 1, 26.
MILLER, E. E., MILLER, R. D. (1956): J. Appl. Phys. *27*, 324.
MILLER, F. G. (1941): U.S. Bureau of Mines R.I. 3595.
MOORE, T. V. (1938): Bull. Amer. Ass. Petrol. Geol. 22, 1237.
MORSE, R. A. (1954): J. Petrol. Technol. No. 12, 42.
MORSE, R. A., et al. (1947a): Producers Monthly *11*, No. 10, 19.
MORSE, R. A. et al. (1947b): Oil and Gas J. *46*, No. 16, 109.
MORSE, R. A., YUSTER, S. T. (1946): Producers Monthly *11*, No. 2, 19.
—— (1947): Oil Weekly *125*, No. 6, 36.
MUSKAT, M. (1934): Physics *5*, 250.
—— (1937): See Introduction.
—— (1945): J. Appl. Phys. *16*, 147.
—— (1946): Producers Monthly *10*, No. 4, 23.

MUSKAT, M. (1947): Trans. AIME *170*, 81.
——— (1948): Producers Monthly *12*, No. 5, 14.
——— (1949a): Oil and Gas J. *47*, No. 42, 121.
——— (1949b): Oil and Gas J. *47*, No. 48, 89.
——— (1949c): The physical principles of oil production. McGraw-Hill, New York.
——— (1950): Trans. AIME *189*, 349.
MUSKAT, M., MERES, M. W. (1936): Physics *7*, 346.
MUSKAT, M., TAYLOR, M. O. (1946): Trans. AIME *165*, 78.
MUSKAT, M., WYCKOFF, R. D. (1934): Trans. AIME *107*, 62.
——— (1935): Trans. AIME *114*, 144.
MUSKAT, M. et al. (1937): Trans. AIME *123*, 69.
MUSKAT, M. et al. (1953): Oil and Gas J. *52*, No. 28, 238.
NIELSEN, R. F. (1949): Producers Monthly *14*, No. 2, 29.
NIELSEN, R. F., MENZIE, D. E. (1948): Producers Monthly *13*, No. 1, 36.
——— (1949): World Oil *129*, No. 1, 144.
NIELSEN, R. F., YUSTER, S. T. (1945): Petrol. Eng. *17*, No. 1.
O'BANNON, L. S. (1924): J. Amer. Soc. Heating Vent, Eng. *30*, 157.
O'CONNOR, G. V. (1946): Chem. Eng. *53*, No. 11, 162.
OFFERINGA, J., VAN DER POEL, C. (1954): Trans. AIME *201*, 310.
OJEDA, E. et al. (1953): Bull. M.I. Exp. Sta. Penn. State Coll. *62*, G-18.
OSOBA, J. S. et al. (1951): Trans. AIME *192*, 47.
PATTON, E. C. (1947): Trans. AIME *170*, 112.
PEIRCE, F. T. et al. (1945): J. Textile Inst. *36*, T169.
PETERS, W. A. (1922): Ind. Eng. Chem. *14*, 476.
PFALZNER, P. M. (1950): See Part VII.
PFISTER, R. J. (1947): Producers Monthly *11*, No. 11, 10.
PFISTER, R. J., BRESTON, J. N. (1947): Producers Monthly *12*, No. 2, 10.
PILLSBURY, A. F., APPLEMAN, D. (1945): Soil Sci. *59*, 115.
PIRET, E. L. et al. (1940): Ind. Eng. Chem. *32*, 861.
PIRSON. S. J. (1946): Oil Weekly *122*, Sept. 9, 45.
——— (1950): Elements of Oil Reservoir Engineering. McGraw-Hill, New York (441 pp.).
PIRVERDYAN, A. M. (1952): Prikl. Mat. Mekh. *16*, 711.
PLUMMER, F. B. et al. (1937): A.P.I. Drill. Prod. Pract., 417.
PLUMMER, F. B., LIVINGSTON, H. K. (1940): Bull. Amer. Ass. Petrol. Geol. *24*, 2163.
PLUMMER, F. B., WOODWARD, J. S. (1937): Trans. AIME *123*, 120.
POLUBARINOVA-KOCHINA, P. YA. (1943): Prikl. Mat. Mekh. *7*, 361.
PORKHAEV, A. P. (1949): Kolloid Zhur. *11*, 346.
PURCELL, W. R. (1950): Trans. AIME *189*, 369.
RAPOPORT, L. A. (1954): AIME T.P. 415-G.
RAPOPORT, L. A., LEAS, W. J. (1951): Trans. AIME *192*, 83.
——— (1953): Trans. AIME *198*, 139.
REID, L. S., HUNTINGTON, R. L. (1938): AIME T.P. 873.
RICHARDS, L. A., MOORE, D. C. (1952): Trans. Amer. Geophys. Un. *33*, 531.
RICHARDSON, J. G., et al. (1952): Trans. AIME *195*, 187.
ROSE, W. (1948): A.P.I. Drill. Prod. Pract., 209.
——— (1949): Trans. AIME *186*, 111.
——— (1951a): Proc. 3rd. World Petrol Congr. *2*, 446.
——— (1951b): Trans. AIME *192*, 373.
——— (1954): Petrol. Eng. 26, No. 4, B58.

ROSE, W., BRUCE, W. A. (1949): Trans. AIME 186, 127.
ROSE, W., WYLLIE, M. R. J. (1949): Trans. AIME 186, 329.
RUSSELL, R. G., et al. (1947): Trans. AIME 170, 51.
RUSSELL, W. L. (1932): New York State Museum Circ. No. 8.
RYDER, H. M. (1947): Producers Monthly 11, No. 7, 13.
────── (1948): World Oil 128, No. 2, 142.
────── (1949): Producers Monthly 13, No. 9, 20.
SAKIADIS, B. C., JOHNSON, A. I. (1954): Ind. Eng. Chem. 46, 1229.
SARCHET, B. (1942): Trans. Amer. Inst. Chem. Eng. 38, 283.
SAYRE, A. T., YUSTER, S. T. (1949): Producers Monthly 13, No. 12, 19.
SCHIFFMAN, L., BRESTON, J. N. (1950): Producers Monthly 14, No. 3, 11.
SCOTT, P. H., ROSE, W. (1953): Trans. AIME 198, 323.
SEN-GUPTA, N. C. (1943): Indian J. Phys. 17, 338.
SHERWOOD, T. K. (1937): Adsorption and extraction. McGraw-Hill, New York.
SIMMONS, C. W., OSBORN, H. B. (1934): Ind. Eng. Chem. 26, 529.
SLOBOD, R. L., CAUDLE, B. H. (1952): Trans. AIME 195, 265.
SPENCER, O. (1949): Secondary Recovery of Oil. Penn. State Coll. Ext. Serv.
SQUIRES, F. (1947): World Oil 127, No. 6, 145.
STAHL, C. D., NIELSEN, R. F. (1951): Producers Monthly 15, No. 10, 25.
STAHL, R. E., HUNTINGTON, R. L. (1943): Trans. AIME 151, 138.
STILES, W. E. (1949): Trans. AIME 186, 9.
SUDER, F. E., CALHOUN, J. C. (1949): A.P.I. Drill. Prod. Pract., 260.
TARNER, J. (1944): Oil Weekly 114, No. 2, 32.
TEMPLETON, C. C. (1953): AIME T.P. 307-G; Trans. AIME 201, 162.
────── (1954): Bull, Amer. Phys. Soc. 29, No. 2, 16.
TERWILLIGER, P. L. et al. Trans. AIME 192, 285.
TERWILLIGER, P. L., YUSTER, S. T. (1946): Producers Monthly 11, No. 1, 42.
────── (1947): Oil Weekly 126, No. 1, 54.
TORREY, P. D. (1943): Oil and Gas J. 1943, Nov. 11, 210.
UCHIDA, M. (1937): J. Japan. Ceram. Ass. 45, 133.
UCHIDA, S., FUJITA, S. (1936): J. Soc. Chem. Ind. Japan 39, 886.
────── (1937): J. Soc. Chem. Ind. Japan 40, 238.
UREN, L. C., BRADSHAW, E. J. (1932): Trans. AIME 98, 438.
UREN, L. C., DOMERECQ, M. (1937): Trans. AIME 114, 25.
UREN, L. C., FAHMY, E. H. (1927): AIME Petrol Dev. and Tech., 318.
USTINOV, N. (1938): Azerbaidzhanskoe Neftyanoe Khoz. 18, No. 4, 18.
VAN WINGEN, H. (1938): Oil Weekly. Oct. 10.
VAN WINGEN, N., JOHNSTON, N. (1946): Oil Weekly 122, No. 5, 24.
VERSCHOOR, H. (1938): Trans. Inst. Chem. Eng. 16, 66.
VERSLUYS, J. (1917): Int. Mitt. Bodenk. 7, 117.
────── (1931): Bull. Amer. Ass. Petrol. Geol. 15, 189.
VILBRANDT, F. C. et al. (1938): Trans. Amer. Inst. Chem. Eng. 34, 51.
VON ROSENBERG, D. U. (1956): See Part II.
WEISMAN, J., BONILLA, C. F. (1950): Ind. Eng. Chem. 42, 1099.
WELGE, H. (1949): Trans. AIME 179, 33.
────── (1952): Trans. AIME 195, 91.
WENTWORTH, C. K. (1947): Pacific Sci. 1, 172.
────── (1951): Proc. Ass. Gén. Bruxelles, Ass. Hyd. Int. (UGGI) 2, 238.
WHITE, A. M. (1935): Trans. Amer. Inst. Chem. Eng. 31, 390.
WHITING, L. L., GUERRERO, E. T. (1951): Oil and Gas J. 50, No. 12, 272.

WILSON, D. A., CALHOUN, J. C. (1952): Oil and Gas J. *51*, No. 3, 175.
WILSON, J. W. (1956): J. Amer. Inst. Chem. Eng. *2*, 94.
WINK, W. A., DEARTH, L. R. (1949): TAPPI *32*, 232.
WYCKOFF, R. D., BOTSET, H. G. (1934): Physics *5*, 265.
——— (1936): Physics *7*, 325.
WYCKOFF, R. D. *et al.* (1933): Trans. AIME *103*, 219.
WYLLIE, M. R. J. (1951): Trans. AIME *192*, 381.
WYLLIE, M. R. J., SPANGLER, M. B. (1952): Bull. Amer. Ass. Petrol. Geol. *36*, 359.
YUSTER, S. T. (1945): J. Inst. Petrol. *31*, 121.
——— (1946): Oil Weekly *121*, No. 8. 36.
——— (1951): Proc. 3rd World Petrol. Congr. *2*, 436.
YUSTER, S. T., CALHOUN, J. C. (1944): Producers Monthly *9*, No. 1.
——— (1945): Producers Monthly *9*, No. 4, 16.
ZENZ, F. A. (1947): Trans. Amer. Inst. Chem. Eng. *43*, 415.

INDEX

Adiabatic expansion of gas, 24
Adsorption, 28, 36 ff.; flow with, 148 ff.; in multiphase flow, 197
Adsorption isotherm, 28, 29, 36 ff.
Adsorption potential, 37
Adzumi constant, 27, 144
Adzumi equation, 27, 144
Agar-Agar, permeability of, 68
Analogue, electrical, 88
Analogy: of flow with other phenomena, 75 ff.,; of laminar flow in porous media with turbulent flow, 120
Angularity, 19, 91
Anisotropic porous media, 63, 171
Area; see surface

Berl saddles: permeability, 68; porosity, 12; specific surface, 14
BET adsorption equation, 38 ff.
Boundary conditions, 24
Boundary effect, 126
Boundary layer theory, 25
Boyle-Mariotte law, 9
Breakthrough saturation, 165, 166
Brick: permeability, 68; porosity, 12
Brinkman theory, 111
Buckley-Leverett case, 163 ff.
Buckley-Leverett equation, 164

Capillaric models, 91 ff., 135, 182
Capillaries, 5
Capillarity, 36, 41
Capillary condensation, 36, 41 ff., 150; in multiphase flow, 197
Capillary potential, 150
Capillary pressure, 43 ff.; correlation with permeability, 90; curve, 49, 50
Capillary rise, 46, 47
Catalyst: adsorption isotherms, 29; porosity, 12; specific surface, 14
Cauchy-Riemann equations, 74
Caverns, 5
Cementation factor, 19
Central limit theorem, 115
Channel closure, 195
Characteristic, of differential equation, 164
Characteristic diffusion time, 198, 202
Cigarette, permeability of, 68
Clay: plasticity of, 21; swelling, 126
Coal, porosity of, 12
Compaction factor, 19
Compressibility, 62; of fluids, 24; of porous media, 20, 22, 82
Compressible porous media, 62, 82
Concentration, of solute, 32
Concentration front, 35

Concrete, porosity of, 12
Connate water, 49
Consolidation, 21
Constraints, 173 ff.
Contact angle, 29 ff.
Continuity condition, 23, 62
Continuous matter theory of fluids, 23
Convection, 31
Cork board, permeability of, 68
Correlation tensor, 121
Correlations: friction factor versus Reynolds number, 128 ff.; high velocity flow, 127
Critical holdup, 194
Critical points, in tower operation, 191

Dam, seepage through, 73, 81
Darcy (unit), 57
Darcy's law, 54 ff., 56 ff.; and anisotropic porous media, 63; and compressible porous media, 62 ff.; immiscible phases, 154; and isotropic porous media, 60; limitations of, 123, 190; miscible phases, 198; and multiphase flow, 154; solutions of, 70 ff., 163 ff.
Deformation, of porous media, 20
Density, 25
Desorption; see adsorption
Diffusion, 148 ff.; molecular, 31 ff., 118
Diffusivity coefficient, 33
Diffusivity equations, 33, 148
Dimensional analysis, of turbulent flow, 133
Directional permeability, 64
Disorder hypothesis, 114
Dispersed feed method, 157, 160
Dispersed porous media, 6
Dispersion, 114 ff.
Dispersivity, 118
Displacement, miscible, 197; molecular, 199; quasistatic, 36, 43 ff.; studies of experimental, 180
Displacement pressure, 49
Dissolution (chemical), 148
Dolomite, porosity of, 12
Drag force, 109
Drag theory, 108 ff.; laminar flow, 108; slip flow, 146; turbulent flow, 138, 139
Drainage, 73, 81
Dupré equation, 29
Dupuit-Forchheimer assumption, 92, 135
Dupuit-Forchheimer theory, 80
Dynamic holdup, 194
Dynamical dispersivity, 118
Dynamical similarity, 176 ff.

Eddies, 25
Effective pore space, 5

Index

Effective porosity, 6
Effective stress, 20
Elasticity, of porous media, 20 ff.
Electricity, flow of, 75
Electro-osmosis, 127
Equivalent pressure, 147
Ergodic hypothesis, 114
Error function, 35
Euler equation, 25

Factor of dispersion, 116
Felt, permeability of, 68
Ferrandon's theory, 63
Fibre glass: permeability, 68; porosity, 12; specific surface, 14
Fibres, flow through, 108
Filter velocity, 60
Filtration: equation, 69; theory, 68
Flooding experiments, 180
Flooding point, 191
Flow: analogy with other phenomena, 75 ff.; of electricity, 75; equations, 123; high rates of, 123; high velocities of, 127
Fluids, 23 ff.
Force potential, 60, 156
Force spaces, 6
Forchheimer equations, 127
Free molecular path, 26, 27
Free surface, 77
Friction factor, 123 ff.; in multiphase flow, 193; modified, 132
Funicular régime, 44

Gas, 24; flow of, 85, 86
Gas drive method, 157, 159
Gas-flooding experiments, 180
Gauss curve, 7
Gauss distribution, 115
Geometrical dispersivity, 118
Geometry of numbers, 18
Ghyben-Herzberg principle, 200
Gibbs equation, 37
Grain diameter, 15; effect on permeability, 90
Grain size, 7, 15, 19
Grain size distribution, 7, 15; correlation with permeability, 90; determination, 15, 16
Gravity flow, 77 ff., 86; graphical method for, 79
Gravity vector, 60

Hafford technique, 157, 159
Hagen-Poiseuille equation, 25, 135
Harkins-Jura equation, 39, 40
Hassler method, 157, 159
Heat: conductivity equation, 83; flow, 76
Heat of adsorption, 28
Heat of wetting, 52
Hodograph transformation, 78
Holdup, 192 ff.
Hooke's law, 20
Hugoniot equation, 165

Hydraulic radius, 46, 99, 100, 123; theories of: laminar flow, 99 ff., multiphase flow, 185 ff.
Hydrostatics, 36 ff.
Hysteresis: adsorption, 41; contact angle, 30

"Ideal" fluid, 24
Imbibition, 171 ff.
Initial conditions, 24
Ink bottle effect, 42
Interconnection of pores, 5, 6
Interfacial tension, 28 ff., 41, 44 ff.
Interstices, molecular, 5
Invasive displacement, 199
Ionic effect, 126
Irrigation, 73 ff., 81
Isotherm; see adsorption isotherm
Isothermal expansion of gas, 24

J curve, 47
Jamin effect, 154

Kaolin, plasticity of, 21
Kelvin equation, 41 ff.
Klinkenberg correction: in multiphase flow, 197; in permeability measurements, 152
Klinkenberg equation, 145, 146
Knudsen equation, 27
Knudsen flow, 125
Knudsen's law, 144
Kozeny-Carman equation, 104
Kozeny constant, 103, 104
Kozeny equation: and multiphase flow, 185 ff.; and single phase flow, 93, 95, 100 ff.
Kozeny theory: and laminar flow, 100 ff.; criticism of, 107; and multiphase flow, 185 ff.; and turbulent flow, 137

Laminar flow, 25; of immiscible fluids, 154 ff.
Langmuir equation, 38, 39
Laplace equation, 3, 71, 75
Laplace operator, 71
Laplace transformation, 84
Leather: permeability, 68; porosity, 12; specific surface, 14
Leverett function, 47, 48
Limestone: permeability, 68; porosity, 12
Linear approximation, to unsteady state flow, 82 ff.
Loading point, 191

Meniscus, 41; curvature of, 46
Mercury injection, 51
Micropores, 5
Microporous media, 148 ff.
Microscopic theory: of fluids, 23; of miscible displacement, 200
Miscible displacement, 197 ff.
Miscible fluids, 31 ff.

Index

Miscibility, 31
Models, of porous media, 19, 91 ff.
Molecular effects, and flow, 125, 144, 197
Molecular flow, 26 ff.
Molecular streaming, 125, 144
Molecular theory of fluids, 23
Multilayer adsorption theory, 38, 39
Multiple phase flow, 153 ff., 167 ff.

Natural packing, 19
Navier-Stokes equation, 24, 111
Neutral stress, 20
Newton's law of motion, 23

Ordered porous media, 6
Orientation: factor, 105; of grains, 91
Overburden pressure, 20 ff.; effect on permeability, 58, 162, 163

Packing, of grains, 16 ff.
Parallel type models, 94
Pendular régime, 43, 44
Pennsylvania State method, 157
Permeability, 57 ff.; anisotropy of, 59; axes of, 64; corrections due to flow anomalies, 152; dependence on overburden pressure, 58; directional, 64; ellipsoid, 64 ff.; empirical correlations, 89 ff.; measurement, 66; physical aspects, 89 ff.; tensor, 63; theories, 89 ff.; units, 57; values, 68; variation in drag theory, 110
Permeability-porosity relationship, 89
Plasticity, of porous media, 21
Point B, 40
Polanyi's theory, 36
Pore, 5
Pore diameter, 6, 7, 14; determination of, 43; effect on permeability, 90
Pore size, 4, 6; determination of, 107
Pore size distribution, 6, 7, 14, 15, 16, 42, 49 ff.; cumulative, 7, 15, 16; curve of, 16; determination of, 14, 42; effect on permeability, 90, 94
Pore space, 5
Pore structure, 6; determination of, 106, 152; effect on permeability, 90
Pore velocity, 92
Porosity, 6; influence on friction factor, 129 ff.; measurement of, 8 ff.; values, 12
Porosity factor, 100
Porosity function, 132
Porosity-permeability relationship, 89
Porous materials, examples of, 5
Powders, 152
Prandtl's mixing length, 126
Pressure, 20
Principal axes (of permeability), 64
Probability, 114 ff.
Probability distribution, 114

Quasistatic displacement, 36, 43 ff.

Radial encroachment, 169
Radial flow equation, 71
Raschig rings: porosity, 12; specific surface, 14
Relative permeability, 155; curves of, 161; determination of, 157 ff.; theories of, 181 ff.
Relative pressure, 38, 41
Relative wettability, 52
Reservoir, 84
Reynolds number, 36, 109, 123 ff., 135; critical, 26, 124 ff., 137, 140; modified, 131; in multiphase flow, 193
Rheological condition: for fluids, 24; for porous media, 20 ff.
Rhombohedral packing, 17, 18
Rock, porosity of, 12
Roughness, of surface, 30
Roundness, 91

Sand: permeability, 68; porosity, 12; specific surface, 14
Sandstone: permeability, 68; porosity, 12
Saturation, 44
Saturation bank, 165
Saturation front, 201
Saturation régime, 43, 44
Scaling: single phase flow, 77; multiphase flow, 176 ff.
Scaling coefficient, 177
Scaling factor, microscopic, 179
Seepage: surface of, 77; velocity, see filter velocity
Serial type model, 95
Shape factor, 131, 145
Shchelkachev's theory, 62 ff.
Shock condition, 165
Shock front, 165
Silica grains, porosity of, 12
Silica powder: permeability, 68; porosity, 12; specific surface, 14
Similarity, microscopic, 178 ff.
Single sample dynamic method, 157, 158
Slate powder: permeability, 68; porosity, 12; specific surface, 14
Slip correction, 144 ff.
Slip flow, 26 ff., 106, 125, 144 ff., 197
Slip term, 144
Soil: permeability, 68; porosity, 12
Specific heat, 24
Specific surface, 6; determination of, 11, 106; effect on permeability, 93 ff.; values, 14
Spherical packings, 12, 16, 17, 44
Sphericity, 132
Stabilized flood, 176 ff.
Stable packing, 18
Stationary liquid method, 157, 158
Statistical models of porous media, 19
Statistical theory: laminar flow, 113; miscible displacement, 201 ff.; slip flow, 146; turbulent flow, 139 ff.
Steady state flow, 71 ff.; liquids, 71; gases, 72

INDEX

Stokes' law, 113
Straight capillaric models, 92
Stream function, 74
Stream line, 74
Stress, in membrane, 76
Stress tensor, 20
Structure determination, 106
Surface (free) in gravity flow, 77
Surface determination, 39, 152
Surface energy, 45
Surface flow, 149, 197
Surface of seepage, 77
Surface tension, 28 ff., 44

Textural constant, 185
Threshold pressure, 154
Tortuosity, 7, 93, 96, 104, 136, 183
Towers, 191 ff.
Transient; *see* unsteady state
Transition point, 25
Tube, flow through, 25

Turbulence, 25, 120, 133 ff., 140 ff., 191 ff.

Unconsolidated porous media, 15
Unsteady state flow, 82 ff., 85 ff.

Velocity: filter-, 60; pore-, 92
Velocity correlation, 121
Velocity potential, 60
Viscosity, 24
Voids, 5
Volume force, 25
Vugs, 15

Water-flooding experiments, 180
Waves, in porous media, 84
Well, flow into, 72, 82, 143, 171
Wettability, 31, 36, 52
Wire crimps: permeability, 68; porosity, 12; specific surface, 14

Yuster effect, 191